GOING
WILD

D1026071

UCD WOMEN'S CENTER

GOING WILD

➤➤➤ ⬅⬅⬅

*Hunting, Animal Rights,
and the Contested
Meaning of Nature*

➤➤➤ ⬅⬅⬅

REVISED AND EXPANDED EDITION

Jan E. Dizard

University of Massachusetts Press

Amherst

Copyright © 1999 by
The University of Massachusetts Press
All rights reserved
Printed in the United States of America
LC
ISBN 1-55849-190-2
Designed by Jack Harrison
Set in Adobe Caslon by Keystone Typesetting, Inc.

Library of Congress Cataloging-in-Publication Data

Dizard, Jan E., 1940–
Going wild : hunting, animal rights, and the contested meaning of nature /
Jan E. Dizard — Rev. and expanded ed.
p.   cm.
Includes bibliographical references (p.   ) and index.
ISBN 1–55849–190–2 (pbk. : alk. paper)
1. Animal rights—Massachusetts—Quabbin Reservation.
2. Deer hunting—Moral and ethical aspects—Massachusetts—Quabbin Reservation.
3. Forest management—Moral and ethical aspects—Massachusetts—Quabbin Reservation.
4. Wildlife management—Moral and ethical aspects—Massachusetts—Quabbin Reservation.
I. Title.
HV4765.M4D593   1999
333.95′965—dc21   99–17091
CIP

British Library Cataloguing in Publication data are available.

For Jesse and Seth—
Your questions got me thinking

Contents

Preface to the Revised and Expanded Edition

THE IDEA for this new edition grew out of a number of requests from readers of *Going Wild* who wanted to know "what happened." Nearly a decade has passed since the first deer hunt in 1991, and the forest has responded much as the MDC foresters had predicted: young saplings are popping up and are less subject to the nibbling of deer, though regeneration is still less than the MDC foresters had hoped for. The cumulative impact of deer overpopulation on the reservation might well have been underestimated, much like the size of the herd seems to have been. But all else equal, the fate of the Quabbin watershed is no doubt more secure now than it was ten years ago.

Of course, "all else" is rarely equal. Managing natural resources is growing steadily more difficult, and managing the Quabbin is no exception. For all the grief the MDC took and still takes over the hunt, nibbling deer may well prove more manageable than humans who want to nibble in their own ways. The greatest challenge on the horizon is coming, paradoxically, from the Environmental Protection Agency. The EPA has ruled that federal clean water standards require that a multimillion dollar filtration plant be installed to process the water flowing to Boston. The MDC is contesting the order and the matter is pending in the courts.

Should the EPA prevail (and it is hard to see how, in the long run, it will not), the MDC's major justification for sharply restricting access to the Quabbin will be severely weakened. The MDC has been able to forestall all sorts of requests for access by insisting, with plenty of empirical evidence to back them up, that water quality could be maintained only if human presence were kept to a minimum. Restricting access thus saves the tax payers hundreds of millions of dollars by avoiding the need for filtration. Against this claim, the pleas of mountain bikers, cross-country skiers, kayakers and the host of other outdoor enthusiasts who would dearly love access to the Quabbin have been relatively easy to turn aside. But if the water is going to be filtered and treated by an EPA-mandated facility, the pressures from recreational advocates will quickly intensify.

Access to the Quabbin has become more compelling in recent years because most of the towns that abut or are near the Quabbin are strapped for cash. Industry in the area continues to decline; farming is also declining; and the region is just far enough away, as were the original towns in their day, from centers of economic growth to the east and west to insure pinched municipal budgets. The Quabbin could be a four-season recreational mecca, which would mean an enormous boost to local economies. If the EPA has its way, clean water standards may doom years of enlightened, albeit embattled, stewardship.

In this way, too, the Quabbin has become an exemplification of our national agony over the environment. Deer, trees, water quality, recreational access—how many needs can we simultaneously meet and still leave the generative base of it all sufficiently robust to insure that the next generation of flora and fauna (and humans) will find their needs met? The more closely I inqure, the more conflicted and uncertain I have become. I am uncomfortable with uncertainty, though over the years I've come to a grudging acceptance of it. A number of people have helped me through my discomfort: Elizabeth Colleran and Linda Heubner, who began as students of mine in the Tufts University School of Veterinary Medicine Masters Program in Animals and Public Policy, became friends and my most telling critics; Ben Minteer, John Organ, and Kevin Richardson, friends with whom I share ideas but not nearly enough time together; and Bob Muth and Martha Mather, without whose encouragement I would, likely as not, have let *Going Wild*, like my bird dogs in mid-January, rest. To all I am grateful, not only for the encouragement and bracing criticism, but for the optimism they have helped me sustain in the face of uncertainty.

Preface to the First Edition

THOUGH I didn't know it at the time, this book began one summer nearly twenty years ago. My wife, our two sons, and I took an entire summer off to camp our way from New England to California and back. It was a memorable trip, not least for the fact that the four of us survived a summer together in the close quarters of a Dodge Dart and an ultralight backpacking tent. As one park followed another, though, I found myself paying nearly as much attention to the people as to the natural wonders.

The shock of the first oil crisis had subsided and the highways and parks were filled with vacationers like ourselves. Sort of. Many, like us, were traveling light. There were also plenty of huge trailers and campers sharing the roadways and campsites. One evening in particular stands out, and it marked the start of this project.

I had mistakenly overshot the campgrounds on the Gunnison in Colorado and then stubbornly insisted that we press on rather than turn back. The result was that at day's end there was no alternative but to rent a site at a barren, dusty commercial "kampground." With much grumbling about the accommodations and the spectacular river canyons we were not camped beside, we fell into our well-worn routine of pitching the tent and setting up for our evening meal. Just as we had finished these chores, a mammoth trailer pulled into a berth no more than thirty yards from our campsite.

We watched in disbelief as the occupants began unpacking for their stay. The trailer disgorged two huge folding picnic tables, each larger in square footage than our tent, one for the children, the other for the adults. Next came television sets, one for each table. Two Saint Bernards joined the happy family scene, adding to the considerable bustle around the vehicle. Several refrigerators and a small microwave oven joined the TVs on the tables, and a bicycle and assorted toys and paraphernalia were soon deployed. Muffling the chatter from the television sets, two air conditioners and the generator that kept them and the assorted electric appliances in business hummed steadily. After sunset, the warm glow from our gas lantern was completely washed out by the glare of the halogen bulbs that bathed our neighbors' site.

We were alternately aghast and amused, though our sense of humor disappeared when it became clear that the noise from the generator and air conditioners would persist through the night. Our sleep disrupted, we could not even take solace in watching the stars or listening to the crickets. We might as well have been trying to sleep on a park bench in the city. For days afterward, we speculated about those "campers." We dubbed their vehicle the Queen Mary. We invented biographies for the family and enjoyed joke after joke at their expense. We also talked seriously, for the first time really, about what camping and backpacking meant to us.

Each of us understood our vacation differently, but a common thread was that in varying degrees we each wanted to encounter nature in a direct way. We wanted to put aside as much of what normally protected and insulated us from the elements as we could, testing our endurance and resourcefulness in the process. To be sure, we knew our experience was contrived and sanitized. Our tent was made of space-age fabric and held up by equally high-tech metal poles. We cooked over elegantly compact gas stoves, and when we hiked out from our base camp, we walked on state-of-the-art boots and carried our stuff—a lot of it—in equally refined packs. Though our investment for all this was minuscule compared to what the Queen Mary's crew must have spent, we had nevertheless been at pains to encounter nature on our own terms.

But, we wondered, what on earth had our campsite neighbors thought they were doing? What nature were they out to observe? What could nature possibly mean under the circumstances they had designed for themselves? What, we finally came to ask ourselves, was nature after all? Could we really say that ours was a purer encounter with nature than theirs? And which of us, when all was told, left the greater imprint—the four of us adding to the already heavy pressure on the backcountry or the crew of the Queen Mary, who probably never strayed more than a hundred yards from asphalt?

We had set out that summer confidently thinking that we knew what nature was and how best to savor it. We returned unsettled and agitated. Nature no longer seemed a fixed idea. It meant different things to different people and, as important, there was no clear way to rank these meanings on some scale of correctness or truth. Moreover, the lines of cars at park entrances, and the wait for backcountry passes, not to mention the obvious signs of erosion that countless vibram-soled boots like ours had caused, made it apparent that we could not experience nature without thereby changing it. The closer we got, the more it changed, both imaginatively and physically.

These were not the sorts of matters I was prepared to handle, either personally or professionally. In the years that followed that trip, I continued to teach and write on the subjects in which I had been trained. But the questions that had arisen that summer kept nagging at me. I began to read more natural history and pay closer attention to the growing nontechnical literature on nature and ecology.

For a while, I loosely organized the literature I encountered as a record of accumulating knowledge and insight. Implicit in this way of proceeding was the comforting notion that as flawed assumptions and mistaken beliefs get stripped away by research, we come closer and closer to knowing what nature is really about. I cannot say exactly when or why my faith in this view began to collapse. Certainly, reading Thomas Kuhn's *The Structure of Scientific Revolutions* had something to do with my shifting perspective. Kuhn argues compellingly that scientific truths are established by agreement among practicing scientists. However much these agreements involve physical evidence, they are also irreducibly based on social conventions. And like all social conventions, they involve culturally mandated biases, distortions, and blind spots. The more I read, the more struck I became by how various are our ideas about nature and how hard—perhaps even impossible—it is to decide among the various claims. What seems settled and indisputable one moment dissolves into ambiguity the next.

In important ways, nature is what we make it out to be. Though in one sense nature exists independently of us, its meaning for us is entirely of our own making. This is not to say that nature does not have qualities that we can apprehend, measure, record, and catalogue. It is to say that the meaning we attach to these qualities matters at least as much as the qualities themselves.

Such thoughts as these probably would have remained unfocused had I not read an article in the local paper three years ago about a controversy brewing over the deer at the Quabbin, a large reservoir and watershed near where I live. The proverbial light bulb flashed. Here, right next door, was an instance of people arguing over precisely the sorts of issues that had become my own preoccupation. I couldn't resolve the quandaries I had been laboring with alone. Perhaps I could get help by studying how others were wrestling with these matters.

The questions remain. I was not so rash as to think that I would settle matters as complex as the meaning of nature or determine the appropriate relationship we should maintain with the natural world. I do hope, though, that *Going Wild* helps to clarify some of the central questions with which we who profess to care about the environment must grapple. Most of all, I hope that *Going Wild* will sharpen the debate over how best to relate to the environment.

Many people have helped me in my endeavors. I especially want to record my huge debt to the men and women who allowed me into their homes and offices and who freely shared with me their thoughts and feelings. Without their cooperation, it would have been impossible to proceed. Regrettably, I cannot acknowledge each of these people by name since a number of them consented to be interviewed with the promise of anonymity. Feelings, as we will see, ran high, and many people worried lest their views offend employers or others with whom they had association. To conceal their identities, I have given them pseudonyms. Ideally, I would have preferred to do this for all the people I interviewed, but this turned out to be inadvisable.

When the idea for this project first struck me, I attended and recorded the series of public hearings that are described in chapter 1. As I reflected on what I had heard and seen at these gatherings, I realized that though I could easily identify the broad divisions that separated people, only by interviewing the protagonists could I arrive at some sense of the reasoning people used in coming to the positions they held. Since a number of the people I would need to interview were officials or had otherwise been publicly involved in the policy-making process, the only way their identities could be concealed would be to change everything—including the name and location of the reservoir. That would have entailed giving up much if not all of the rich specificity of the Quabbin. Instead, I asked those people whose identities I could not easily mask to permit me to use their real names and, where relevant, their organizational affiliation. As a consequence, some of the names used in *Going Wild* are real and some are pseudonyms. I indicate which are which in the "Sources" that appear after chapter 6.

I should also note that I chose the people I interviewed for different reasons. Some were selected because they were central to the controversy—policy-makers, leaders of the opposition, spokespersons for organizations involved in the policy process. Others came to my attention through their comments at the public meetings or letters they wrote to the editors of local newspapers. My intention was to make sure that every major shade of opinion, pro and con, was represented. I don't mean to suggest that the people I interviewed are representative of those holding a particular point of view. Though I feel confident that the views expressed by the people I interviewed resonate broadly and deeply, mine is not a sampling of opinion that permits formal generalization to the population as a whole or even to all those who are interested in the fate of the Quabbin. My goal in *Going Wild* is to explore people's ideas and positions, not their motives or personalities.

Overall, I met with extraordinary cooperation. Only four people I contacted declined to be interviewed, three of them opponents of the hunt, two of them advocates of animal rights. One of the latter said that she did not grant interviews to anyone not in the animal liberation movement. Another opponent, a person who had read columns I had written in the local press on hunting and fishing, refused saying that she did not think that someone who hunts would present her views objectively. But aside from these refusals, everyone whom I interviewed was forthcoming and interested in the project. I cannot thank them enough.

This said, I want especially to thank Robert O'Connor. Not only did he consent to be interviewed, but he also gave me permission to interview members of his staff, made available to me reports and studies he and the Metropolitan District Commission had conducted or commissioned, and, going the last mile, carefully read an early draft of the manuscript, providing me with details that I otherwise would have missed or distorted. Though he had much at stake in the outcome of this study, he did not in any way try to dictate what I should say or what conclusions I should draw. I admire his candor and integrity. Without his cooperation, this study simply could not have been done.

When the study was in its earliest, unfocused stages, Elisa Campbell and Eva Schiffer, each of whom had been following the "deer problem" at the Quabbin long before I got wind of it, met and talked with me and availed me of their considerable store of information. Eva kindly shared with me all the reports and correspondence she had gathered. Their generosity saved me untold hours I would otherwise have had to spend tracking down documents and background reports.

My friend Tom Looker helped in ways almost too numerous to mention. When he invited me to join him in teaching a course he and Dick Schmalz, an Amherst College professor of fine arts, had instituted, a course titled "The Imagined Landscape," we embarked upon an intellectual enterprise that I have drawn upon at every step of the way in writing *Going Wild*. Our conversations over the years during which the book was taking shape helped me achieve a clearer sense of my own imagined landscape and the subtleties of imagination itself.

Several friends have read all of an earlier version of *Going Wild*. Their combined efforts have done much to make *Going Wild* more accurate, spare, and exacting. Without their help, much would have been left murky, and some things would have been just plain wrong. Richard DeGraaf, wildlife biologist with the U.S. Forest Service, was painstaking with the manuscript, and our conversations while pursuing upland birds enabled me to see many technical issues more clearly than I would have otherwise. He has been a fount of knowledge and good sense. David Wellman, as he has done so often over our friendship of thirty years, kept me faithful to intellectual traditions of social analysis that demand rigor, objectivity, and tough-mindedness. To Jonathan Tucker goes credit for helping keep my sentences uncluttered and for making the analysis both crisper and tighter than it otherwise would have been. An author could not want three more thorough and constructive critics.

Bruce Spencer offered a number of helpful comments on the first draft of the book. Ali Crolius also read and made many useful comments on the first draft. Though her deepest commitments have led her to reject hunting, she nevertheless was willing to engage *Going Wild* on its own terms. Our discussions were most helpful and her encouragement gratifying. Would that philosophical opponents were always as generous and open as Ali has been with me.

Bill Chaloupka, Margaret Knox, Jane Bennett, and Tom Dumm gave encouragement along the way as the present manuscript expanded from a paper to a book manuscript. Bob Muth, with whom I now am collaborating, along with David Loomis, on a study of the men and women who hunted in the first Quabbin hunt, has been a consistent source of stimulation and support. Colleagues and friends Deborah Gewertz and Fred Errington were also helpful early on and their continued interest in this project has spurred me on.

The trustees of Amherst College have been most generous in their support. Not only did they allow me a semester's leave so that I could devote my full energies to writing, they also awarded me a grant that enabled me to do the interviewing and to hire someone to transcribe what turned out to be nearly two thousand pages of interviews. Not just "someone," let me add. Brenda Hanning labored long and lovingly with the tapes and produced transcriptions with a speed and accuracy that were remarkable. I am also grateful for Nina Barten's careful copyediting.

Finally, I wish to record my enormous good fortune in having met Clark Dougan. His early interest in this project and his patient support, editorial acumen, and encouragement were crucial at every step of the way.

With such assistance as has been generously given me, the research and writing of this book have been more gratifying than anyone would have reason to expect. My only regret, and it is a large one, is that I could not find a way to resolve the differences between the partisans. As I interviewed each person in turn, I found myself drawn this way and that, as much by the sincerity with which the views were held as by the logic of the views themselves. This is more than my personal dilemma. As I hope to make clear in what follows, it is our collective dilemma. While everyone wants a better environment, there is no agreement on what that environment should actually look like, much less on how to achieve it. It might be easier, in fact, to generate the will and technology to restore and protect what remains of our natural heritage than it is to reach a consensus about what that heritage is. Those who would preside over these matters will have to become far more sensitive than they have been to the cultural and social dynamics involved in our interactions with nature. In the end, these human dimensions may be more important than technical virtuosity or scientific mastery.

GOING WILD

1

What's Wild?

Nature is perhaps the most complex word in the language.
RAYMOND WILLIAMS, *Keywords*

HEADLIGHTS PIERCED the darkness, their stark whiteness exaggerating the early winter cold. As vehicles approached and passed one another, their beams set the occupants aglow with the unmistakable blaze orange that designates the wearer a HUNTER. The profusion of blaze orange announced that it was opening day of the Massachusetts nine-day shotgun season for deer. It would be legal to shoot beginning at 6:40 A.M., one half-hour before sunrise, by which time there would be enough light for hunters to reliably make out the shape of a deer and determine whether or not the animal had antlers.

The volume of traffic on Route 202 had begun to mount around four A.M., and by five the highway was crawling with cars. Hunters were intent on getting into the woods and settled into their chosen spots well before legal shooting time. Some had been hunting in these woods for years and others had been scouting the woods for weeks before opening day. They knew precisely where they would park and what log or tree or rock outcropping they would be heading for. Still others were going to make a stab in the dark, hoping that the sheer numbers of hunters in the woods would keep the deer on the move and that they would have a chance for a good shot even though they had not selected a spot beforehand.

This stretch of Route 202 is named for Daniel Shays, who, in the early years of the new republic, led a revolt against the Boston finan-

3

ciers who were pressing hard the farmers of Western Massachusetts. The highway runs north-south along the western edge of the largest state-owned tract of open land in Massachusetts. The land, roughly 55,000 acres, surrounds the Quabbin Reservoir, formed by a dam on the Swift River, and the source of drinking water for most of the greater Boston metropolitan area.

This stretch of the Daniel Shays Highway has been a magnet for hunters for years. Though the forests east of the highway that surround the reservoir have been off limits to hunters since the 1930s, when the work on the dam and the watershed began, the west side of the highway has been continuously open to hunting, except for small private holdings that have been posted. The hunting there has been good. Deer on the west side of the road are regularly supplemented by migrants from the other, protected, side of the road. Hunters come from far and wide each December lured by the prospect of abundant deer.

Residents of the area liken the nine days of deer season to an invasion. Just before the 1991 deer season opened, the town of Pelham, the stronghold of the rebel Shays, enacted an ordinance sharply restricting parking along the township's roadways, and signs had gone up all along the west side of the highway alerting hunters to the new restrictions. The pull-offs that were not restricted quickly filled with pickup trucks, vans, and station wagons.

On this day, December 2, 1991, however, some of the hunters had their parking arrangements all set in advance. Instead of cruising the west side of the road for a place, these hunters were turning east and entering one of several clearly marked roads into the Quabbin Reservation. Once beyond sight of the highway, they came upon uniformed officers of the Metropolitan District Commission (MDC), the state agency responsible for maintaining the watershed and reservoir. The MDC police checked the specially prepared permits each hunter presented. If everything was in order, the hunter was issued a tag to be returned when he or she checked out, whether to leave for the day or simply to take a break from the hunt and catch a quick bite at one of the small eateries located along the highway. The hunters were then directed to a designated parking area. For the first time in over fifty years, the deer of the Quabbin would be legal game.

Locals—no one knows for certain how many—had been illegally shooting deer on the reservation from the beginning. But this was different. This morning, a contingent of 300 licensed hunters, to be followed by two more groups of 300 hunters on subsequent three-day periods of the state's nine-day deer season, would be allowed, indeed assisted, to hunt in an area roughly equal to fifteen percent of the total land mass of the reservation. How could it come to pass that hunters would be invited into one of the most beautiful and wild tracts in southern New England?

The Quabbin as Wilderness

The area now known as the Quabbin was settled by Europeans in the late seventeenth century, and the first town in what came to be called the Swift River Valley was incorporated in 1754. Before the arrival of Europeans, Native Americans had also periodically cleared blocks of land in the valley, using fire rather than steel. They burned in order to clear land for staple crops as well as to create areas of regenerating vegetation that attracted the game animals they depended upon for food and hides. It is not clear how long the natives had been settled in the valley, but there is no reason to doubt that the area had been feeling the impact of the native hunters and horticulturists for many centuries before the arrival of the English. In the narrow sense of the word, the valley has not been a "wilderness," a place unmodified by human activity, for a millennium or more.

Native American labors notwithstanding, the valley must have appeared "wild" to the eyes of the settlers. Used to clearly bounded pastures and fields as well as managed woodlands, they would have viewed the natives' efforts as disorderly and as chaotic as nature itself. They toiled to bring order to the unruliness of nature. As the forest was beaten back to make way for home sites, crop lands, and pasture, they built fences, walls, and roads, and later used the water courses to power small mills. Experts in reconstructing the botanical past are certain that, with the exception of a few small swampy nooks and crannies, nearly every acre that now comprises the Quabbin Reservation was cleared of forest at least twice since it was settled by whites.

By the late nineteenth century, the valley was home to four small towns with an aggregate population of two to three thousand souls. The boisterous growth that was occurring to the east in Lynn, Lowell, Boston, and New Bedford, and to the west in Springfield, Pittsfield, and North Adams, was worlds away to the residents of the valley. Though they were linked to the hustle and bustle around them by a rail line, the valley remained solidly agricultural and substantially self-sufficient. Valley residents no doubt found strength in living out Yankee virtues of thrift and resilient independence. But their isolation and independence made them vulnerable in the end. Had factories and commerce moved in, it is possible that the towns of Enfield, Greenwich, Dana, and Prescott would not now be submerged beneath waters impounded by one of the largest earthen dams in the country.

Early in this century, Boston officials recognized that they would soon have to increase substantially the sources of potable water to meet the needs of an expanding metropolis. As they had done several times earlier, they looked west. The Swift River Valley was easily the most compelling location for a reservoir. The topography was ideal for a large reservoir, and the excavation that would be required to construct the necessary tunnels and aqueducts connecting the future reservoir to existing water storage and distribution networks were well within the budgetary and technical capabilities of the Commonwealth.

The valley's residents had little capacity to resist the designs of water-hungry Boston. With a modest economic base and a small population, their political clout was minimal. Moreover, even a scheme of generous compensation would not break the state's bank because there was almost no heavy capital investment, aside from the land itself, that would have to be bought out. Given these favorable conditions, the coming of the reservoir was all but inevitable. By 1915, rumors of the impending project were circulating and the residents of the valley had begun to make plans.

By all accounts, the residents did not put up much organized resistance. From the ignominious defeat of Daniel Shays on, folks in the western part of the state have had little from which to take encouragement in battles with Boston. One by one, landowners and business people settled upon buy-out terms, and before long, those who were inclined to fight or at least hold out were confronted by the consequences of their neighbors' coming to terms: homes were no longer maintained, fences were left untended; investment, both financial and emotional, quickly ground to a halt; and residents began to leave. The reasons for holding out disappeared. Work began in 1928, and the dam and connecting aqueducts and tunnels were completed eleven years later. It took seven more years for the reservoir to fill up. By the end of World War Two, clear, clean Quabbin water was flowing east to Boston and environs.

To this day, the Quabbin remains one of only a handful of large water supplies that the Environmental Protection Agency has not required to install filtration equipment. There are two reasons for the purity of the water collected behind the Winsor Dam. First and most important, there are no major population or industrial centers near the reservoir nor along the major feeder streams that flow into the reservoir. Secondly, from the moment the last town, Enfield, was officially closed and the last person removed in 1928, the land surrounding the reservoir has been assiduously managed by the MDC to provide as much buffer, filtration, and insulation from human activity as possible.

While work on the dam proceeded, buildings were demolished or moved to new locations off the reserve. Trees and brush were cut and pastures cleaned in what would become the basin for the reservoir. Many of the former residents, idled by the loss of their farms and the Great Depression that held the country in its grip, worked on the dam or in the gangs that prepared the valley for the inundation to come. The hurricane that swept inland in 1938 added more work. The storm blew down large sections of the forested hillsides that would soon become shoreline and watershed. The debris from the storm needed to be cleared to reduce the threat of fire and to cut down the amount of decaying organic matter that might compromise the reservoir once it began to fill.

Except for the work crews, the land that comprised the reservation was declared off limits to the general public. As the water level slowly rose, the forest began to reclaim the homesites and fields. Wildlife populations, no longer having to compete with agriculture and no longer thinned by hunters, prospered on the new growth—the tender shoots and buds, and the plethora of insects and other small prey that thrive in such settings. In order to stabilize the soils of the watershed, the MDC planted a number of red pine stands in areas that had been hit by the hurricane. Though not common in the area, red pine was chosen because it is fast growing and promised commercial value upon reaching maturity seventy years or so after planting.

Aside from the planting of red pines, most management effort in the early years was devoted to regulating public access. Fishing along the shore was allowed for the first time in 1946 after intensive lobbying by sport fishermen. Restricted numbers of fishing boats began to be permitted in the early 1950s. Fishing on the reservoir was enhanced by the stocking of several species of sport fish like lake trout and landlocked salmon as well as forage fish to sustain them. Otherwise, nature was largely left to its own devices.

By the 1950s, the land surrounding the reservoir had become heavily forested, predominantly by oaks. Wildlife was abundant and diverse. With few predators present and with hunting forbidden, many animal populations soared, particularly the population of white-tailed deer. It is thought that the herd peaked by the mid-1950s well in excess of sixty per square mile, or roughly six times to eight times the density that is common outside the reservation. At some point, no one is sure exactly when, beavers took up residence. Long absent from southern New England, somehow they found their way back. Later, coyotes also returned, joining bobcats and fisher cats as the major predators in the Quabbin. In the late 1980s, moose put in occasional appearances, and there is now a distinct possibility that breeding pairs have or will soon establish themselves on the reservation. Reliable sightings of mountain lions began in 1969 and have become more frequent, though there is no evidence that the cat, long thought extinct in the East, has taken up permanent residence in the Quabbin.

This resurgence of flora and fauna made the Quabbin seem more and more like an island of wilderness in one of the nation's most densely settled states. Though dwarfed by the large national forests and parks of the West and by the Adirondack Park in upstate New York, the Quabbin is a very large undeveloped area for Southern New England. The combination of restricted public access, sheer expanse, and dense forest all contributed to the perception that the Quabbin was a vast wilderness, a symbol of what things would have looked like had the Mayflower never landed. The evidence of previous human land use that survived the cleansing efforts of the MDC—the roadbeds that get exposed when water levels drop, the overgrown outlines of stone walls that marked the boundaries of former fields, the cellar holes—only added to the mystique that came to envelop the Quabbin. The writer Thomas Conuel captured the essence of this mystique in the title of his careful appreciation of the Quabbin: "Quabbin: The Accidental Wilderness."[1]

As Conuel knows well, there is considerable irony here, not only in designating the Quabbin "wilderness," but also in attributing it to accident. The Quabbin is, after all, completely contrived, much of it with the benefit of the best science and engineering available. Of course there have been "accidents": when the plans were laid for the reservoir and its protective watershed, no one could have anticipated the return of beaver or coyotes. These were "gifts of nature." But such unintended or accidental emblems of wildness, coupled with sharp restrictions of human activity, made the Quabbin seem especially wild, and also masked the Metropolitan District Commission's imprint.

While early management efforts were narrowly focused on stabilizing the watershed with forest cover and discouraging public access, it was not long before MDC's management activities grew more complex and ambitious. The red pines began to become a problem. As they grew, they crowded out everything beneath them. More worrisome, though, was the fire hazard they posed. The groves had been closely planted and required thinning. Many were planted on sites that turned out to be not particularly well suited to them. Moreover, by the time the pines were reaching harvestable age, their commercial value had declined in response to shifting tastes in the wood products industry. This meant that the MDC was faced with the need to maintain the pines without hope of offsetting the costs by the sale of the timber.

Fire suppression, a concern from the beginning, became more of an issue as the forest aged and limbs and trees got diseased and died. Even though restricted public access reduced some of the risk of fires, careless visitors had started fires and there was always the possibility of lightning igniting a blaze. The MDC had been assiduous in its efforts to reduce major sources of fuel as much as possible—dead limbs, diseased trees, anything that might set off a conflagration or accelerate it should one get going; but with some 55,000 acres of mostly forested land, this was no simple task.

Fire is a threat because it destroys ground cover, exposing the soil to erosion, the runoff from which carries a heavy load of decaying organic matter into the reservoir. After years of study and close examination of the experiences in other watersheds, the Commission came to the conclusion that the most "fireproof" forest cover would be a a wide mix of tree species as well as a wide range of age classes of those species. A homogeneous forest would not be problematic in its early stages, but as it matured, die-offs would come in large batches and would present the potential for an enormous amount of fuel all at once. Moreover, a homogeneous forest is far more susceptible to devastation from disease, insect attack, or storm.

As the forest that had "volunteered" grew to fill the vacuum created when the MDC and the hurricane of 1938 collaborated in clearing much of the land that was to surround and protect the reservoir, the Commission began to think of ways to shape the forest to ensure the integrity of the water supply. In the early 1960s, MDC foresters carried out the first systematic inventory of trees on the reservation. This inventory provided a base line from which changes could be measured and assessed. Among the many things that the inventory revealed was a surprisingly low rate of regeneration, particularly of oak, the dominant species on the reservation. Several reasons were offered, including the possibility that the large deer herd, exercising its preference for oak, was eating an inordinate amount of the new growth. At the time, though, there was no sense of alarm, much less of urgency, attached to this finding. The forest was still young and healthy and quite diverse, possibly as healthy and diverse as any in New England. And good, pure water was flowing. Well, almost.

Boston's water distribution network was ancient and full of leaks. Consumer awareness of the advantages of water conservation was virtually nil. The population and economic activity in the greater metropolitan area were growing, and the demand for the Quabbin's water was rising. This increase in demand coincided with a succession of years of below average rainfall, first in the 1960s and then, after the reservoir reached its highest elevation to date in 1984, in the late 1980s. As the water level of the huge impoundment fell, it exposed a ghostly tableau of cellar holes, stone walls, and old roadbeds stretching out from the expanding shoreline. Boston's water officials voiced concern and, following long habit, looked westward for a way to augment the reserves of the Quabbin. They fastened their sights on the Connecticut River and began to talk of diverting some of its flow into the reservoir.

Accustomed to acquiessence from residents in the hinterland, officials were taken aback by the fury unleashed by the suggestion of diversion. By the 1980s, environmentalism had become a household word, and the communities adjacent to the Quabbin were in the vanguard of this new consciousness. Rapid expansion of the state university in nearby Amherst had drawn thousands of students, and hundreds of teachers, researchers, and technicians to the area adjacent to the Quabbin, and they were not about to sit passively by while remote bureaucrats ran roughshod over the pastoral landscape.

To make a long and interesting story short, the diversion was blocked. With no new water source ready to hand, Boston water officials launched a major water conservation campaign aimed at reducing consumption. They also made major investments in patching up the leaks in the distribution system. For its part, the MDC searched for ways of increasing the amount of water flowing into the reservoir. One way was to thin the forest cover, thereby increasing the runoff. This seemed a good idea, especially since thinning also assisted with fire suppression. Besides, there were those red pine plantations to be dealt with.

In the early seventies, just after the first water crisis, when the reservoir was still well below capacity, several of the red pine groves were clear-cut and a more general but very selective thinning was carried out. Water yields edged up. At the same time, consumption dropped even more notably, and more nearly normal rainfalls returned. The reservoir began to refill, and the skeletal remains of the old villages slowly were restored to their watery resting places. Crisis ended.

The deer took well to this turn of events. The exposed shoreline burst forth with new growth, and the clear-cuts and thinning added browse. After peaking in the early to mid-fifties, the herd had started to decline, perhaps to as low as twenty-five to thirty per square mile. This was still more than was common outside the reservation, and the decline was expected by everyone who studied wildlife. Wildlife biologists had long believed that populations rise sharply and, after exceeding the carrying capacity of their habitat, experience a crash, their numbers falling well below carrying capacity. Slowly, things then return to something of an equilibrium between the food supply, predators, and the species in question.

Whether this scenario would ever have played itself out for the deer of the Quabbin is anyone's guess. Indeed, as we shall see later, experts have begun to question this model of wildlife population dynamics. In any event, after the decline had set in, the region's winters began to moderate, reducing the stress on the deer at the same time that low water levels and the MDC's cutting and thinning increased their food supply. A major population crash never occurred. The deer herd quickly rebounded, and, though it apparently never returned to the peaks of the early fifties, it stabilized at levels four to five times what would be expected elsewhere in the region.

The deer became a kind of totem for the Quabbin. Epitomizing grace, beauty and tranquillity, the deer seemed to symbolize all that was happening at the reservation. Mother Nature was making a comeback. Animal life abounded. Visitors came from far and wide to see the wildlife, especially the deer. With predators at a minimum and hunting forbidden, the deer grew tame. Usually shy and nocturnal, they could be seen at all hours of the day or night, peacefully ambling along, grazing here and there. The Quabbin quickly came to represent an idealized, tame wilderness, a place where all the negative effects of human interaction with wild animals and the environment were absent. Quabbin Park, a small area near the administrative buildings of the MDC set aside for picnickers and the general public, was open around the clock prior to 1970, to accommodate people eager to see nature, as one person I spoke with put it, "as it was meant to be."

Wildlife, but especially the deer, were assets to the MDC. By stressing the wilderness values of the Quabbin, epitomized by the deer, the Commission was building a public constituency that they could call upon for support whenever an interest group pressed for greater public access. Fishing, itself permitted as a result of pressures that the MDC was too weak politically to resist, was used by all sorts of recreational advocates, from those who wanted to sail on the reservoir to those who wanted the reservation opened for cross-country skiing, as an argument for their cause. As the number of visitors steadily rose, the Commission had its hands full trying to hold the line. The deer helped the agency make a case for restrictions on public access that might otherwise have been hard to defend.

The deer were more than a potent symbol for the MDC's public relations. Worries about the adequacy of water supplies made water yield a recurring issue. By grazing the understory, the deer were, in effect, working in tandem with the MDC foresters. By nibbling young shoots and nipping young trees in the bud, the deer were grooming the forest more thoroughly than could possibly be done by humans. Not only did this enhance water yield, it also opened the forest up. Visitors could enjoy broad views of the forest—and more readily see deer. The MDC was managing the watershed, but the enchanting deer and the lush greenery made it easy to overlook MDC work and see only the handiwork of nature.

Management was not restricted to forestry. Not all the wild creatures in the Quabbin had found their way there on their own. As we have already noted, nonnative fish species were stocked in the reservoir. But more spectacularly, the bald eagle was introduced to the Quabbin in the early 1980s. Now people flock to the Quabbin in the late fall and winter for a chance to see the magnificent raptor soar over the reservoir. Especially lucky visitors may watch eagles compete with coyotes for morsels of venison. When the reservoir freezes over, coyotes (and sometimes packs of free-running domestic dogs) will chase deer out onto the ice. On ice, the deer slip and slide, often breaking their hips, and become easy prey for the canines. Eagles are quick to join the feast, and a memorable spectacle unfolds.

The successful establishment of breeding pairs of eagles was the painstaking work of Jack Swedburg, a wildlife photographer employed by the Division of Fisheries and Wildlife, for whom the eagle project became a career-capping mission. The eagles have made the Quabbin seem that much more a wilderness, though again it was scarcely accidental. That the eagle was, in all likelihood, only an occasional visitor to the Swift River Valley before the Quabbin was constructed is an irony largely lost on people captivated by the vision of a wilderness within an hour of a major metropolis. Before the Quabbin, the Swift River was hardly a large enough watercourse to be of more than passing interest to eagles. There were far better fishing grounds only a few wing beats away in any direction. But never mind. The eagles' "return" is inspiring, a dramatic and comforting example of how we can restore what we have so cavalierly taken for granted and come close to losing for good.

Managing a Wilderness: Paradox or Oxymoron?

The very different careers of eagles and deer at the Quabbin may well be defining parables for our time. The great bird, on the endangered species list and certainly a precarious presence on the eastern seaboard, requires heavy support and vigorous management. Deer, by contrast, have been prodigiously adaptive and their numbers are skyrocketing all across their very broad range. Where eagles require support and protection in order to survive, deer require a very different sort of human assistance—they must have their numbers controlled. Most environmentalists applaud efforts to save eagles and increase their numbers and range. Efforts to manage deer are another matter. In the difference can be traced our complex response to nature and our ambivalence toward "wilderness."

Terry Tempest Williams, a writer uncommonly skilled at bringing landscapes to life, defines wilderness as something "raw and self-defined."[2] It is an apt and rigorous definition. Though she did not intend it, it is also a definition that essentially eliminates all but the remotest corners of the earth, and maybe not even these, from qualifying as "wilderness." When scientists discovered an ozone hole over the Antarctic, even that redoubt could not longer be described as self-defined. To be sure, blizzards or hurricanes or volcanoes remain self-defined—they come and go as they please and work what they will, and we adapt as best we can.

But if some things in nature continue to be self-defined, most of creation has not. Our land, fully as much as our landscapes, has long since become what we have made it, intentionally or otherwise. Our forests and our wildlife, too, are what we have made them. In some instances, the sculpting has been highly conscious and motivated, as in the reforestation that logging companies have engaged in after they have swept an area clear of timber, or when, in the early years of this century, the Forest Service and other federal agencies pursued to the last creature a campaign to rid the western range of the wolf. We may not be able to rid the grizzly of its occasional murderous impulses, but we can rid ourselves of the grizzly. They remain on this earth because we, ambivalently, will it so, not because the bear has any say in the matter.

With so little left that is "raw and self-defined," some have bemoaned the "vanishing wilderness" or, in Bill McKibben's eye-catching phrase, the "end of nature."[3] Though I would not for a moment wish to blunt the sense of loss and urgency that McKibben feels, nor deflect from the gravity of the predicament he describes, I think it wrong to say that nature is about to end or that wilderness is no more. We may not like having lost what we have lost and we almost certainly will not like what lies ahead, if McKibben's predictions are even close to the mark. But places like the Quabbin show us just how pliable our conceptions of nature and of wilderness are. And even though it can scarcely be claimed that the Quabbin is either raw or self-defined, it is certainly a place where human control is by no means complete. There is a wildness there, if not actual wilderness, despite all the human contrivance and effort that have made it what it is today.

Fifty miles to the east of the Quabbin, and roughly a hundred years before the Quabbin came into existence, Henry David Thoreau found wildness in, of all places, Concord. Concord had been settled for nearly two hundred years before Thoreau headed for the shores of Walden Pond, there to savor the mysteries of wild nature. What Thoreau meant by "wildness" was obviously not what Terry Tempest Williams means today by "wilderness." Thoreau knew that hardly a square foot of Concord had not, at some time or other, felt the weight of human presence. Rather, Thoreau was asserting that there were wonders in and around the pond that were new and captivating to him, things that he could carefully observe, catalogue, marvel at. More important, they were things that, if not exactly self-defining, nonetheless proceeded by their own lights, ebbing here and flowing there, depending upon circumstances that they, like we, could never fully anticipate or control. From these rhythms, and the ways discrete rhythms blended together into an ensemble, Thoreau drew meaning and inspiration. Nature was a virtually limitless, intricate tableau at one's very doorstep. By close observation of nature, Thoreau suggested, we can learn of these intricacies and, at the same time, come to terms with ourselves. Studying how nature works, how delicately each creature must adjust itself to its milieu, teaches us how we ought to live. Nature study helps us separate fancy and frill from the basics.[4]

Wildness for Thoreau was all that was exterior, all that existed without our agency. Indeed, Thoreau clearly resented some, if not all, human interference. Monkey-wrenching, Edward Abbey's picaresque agenda of sabotage to save nature from ravenous human appetite, occurred to Thoreau when he reflected on how a proposed dam on the Merrimack would stymie spawning salmon. Thoreau is arguably one of the first Americans to assert that humans should stop tampering with and instead begin to learn how to live in harmony with nature. He was contemptuous of the absurd lengths to which his fellow Concordians went in order to insulate themselves from the "real world."

Thoreau's implicit model of nature was one of balance, of things fitted together in complex webs of interactive dependency, and of predictable natural rhythms. Nature, seen in these terms, is benign. Humans should accept their part in this scheme of things rather than continually hurl themselves against nature in vain (in both senses of the word) efforts to improve upon it. By contrast, most of Thoreau's contemporaries, had they thought about it, would not have been inclined to see nature as benign. Instead, they would most likely have thought nature unpredictable and, all in all, harsh and uncaring.

This view, of course, gives rise to the notion that nature and wilderness need to be tamed. Thoreau's countrymen were doing just that, and with a vengeance. With the Puritan William Bradford's words still reverberating—"a howling wilderness" was how he had described coastal Massachusetts—Americans had busied themselves building roads and dams and clearing the land for farms and towns. Thoreau, whom the environmental historian Roderick Nash places among the earliest to raise alarm over the environmental consequences of unbridled development, anticipated the EPA: "If some are prosecuted for abusing children, others deserve to be prosecuted for maltreating the face of nature committed to their care."[5]

While Thoreau deplored the steady degradation of nature at the hands of his countrymen, he nonetheless found wildness all around him. Though he could not convince his contemporaries to treat the land more gently, he could personally move closer to nature. In this, Thoreau might be seen as quintessentially American. The drive to settle the West was at least in part a drive to start afresh, to leave behind the spent soils, fouled waters, and corrupt politics of the settled East for the virgin lands to the west. In the same fashion, our contemporaries have turned their backs on the cities to seek solace in suburbs that promise, if not the wildness of Walden Pond, at least a version of closeness to nature. Believing with Thoreau that nature is benign, we turn to it for solace and restoration.

Once we have made our escape, however, we cannot seem to resist the impulse to alter nature. We no sooner get close to it than we begin to cut back the trees, plant uniform carpets of lawn, wage war on woodchucks, and deploy a panoply of chemicals to hold insects and weeds at bay. Up close, nature presents itself less as a benign, restorative force than as a constant challenge and impediment to our desires. The writer Michael Pollan captures this tension as well as anyone. In his recent book on gardening, *Second Nature*, he recounts coming to terms with the animals with whom he shared his garden.

> When I finally did come back to the garden, I was coming from the city and brought many of the city man's easy ideas about the landscape and its inhabitants. . . . To nuke a garden with insecticide, to level a rifle sight at the back of a woodchuck in flatfooted retreat, to erect an electric barricade around a vegetable patch: such measures, I felt, were excessive, even irresponsible.[6]

But after repeatedly losing lovingly cared for and eagerly anticipated plants to predators, Pollan changes his attitude.

> My early efforts at harmonious design were lost on the surrounding landscape, whose inhabitants promptly sought to take advantage of my naive romanticism. . . . Under this many-fronted assault, it did not take long for most of my easy, liberal attitudes toward the landscape to fall. (42)

We seem to be ambivalent about nature. The traditions we draw upon to make sense of ourselves in relation to nature are rich and complex and far from settled. Being unsettled, they allow considerable room for disagreement, controversy, and bitter division. This means that our thinking on these matters is subject to abrupt about-faces. We lurch from the belief that our science and technology can harness nature to a dread that our actions are killing the planet.

This ambivalence has been played out again and again over the last century and a half as the pace of urbanization and industrialization accelerated, with the environmental consequences that now weigh so heavily on us all. Champions of nature, like Thoreau, have warned us repeatedly of both the material and spiritual loss we face if we persist in our efforts to control and modify nature. A few years after Thoreau's death, George Perkins Marsh, a prominent and widely travelled American diplomat, wrote what might well be the first "environmentalist" text, *Man and Nature; Or, Physical Geography as Modified by Human Action* (1864). Marsh was convinced that all living organisms were held in a delicate and complex balance that we should be loathe to upset. Nash (1989) writes, quoting Marsh:

> Anticipating the ecological perspective of the twentieth century, Marsh warned that the interrelatedness of "animal and vegetable life is too complicated a problem for human intelligence to solve, and we can never know how wide a circle of disturbance we produce in the harmonies of nature when we throw the smallest pebble in the ocean of organic life." (38)

In so delicately balanced an arrangement, the best, the most prudent, thing to do, obviously, is to walk as lightly as possible on the earth. Walking lightly is exactly what John Muir did. He hiked to

Canada, to the Gulf of Mexico, and, of course, up and down the range of the Sierras in California. Of all the early writer/naturalists who helped to promote the idea of a benign and balanced nature, none was as widely read nor as influential as Muir. He wrote rhapsodically of nature and its wonders. Readers accustomed to bloodcurdling tales of encounters with grizzlies were no doubt surprised to read Muir's accounts of the great bears' frolics and horseplay. They were probably even more impressed that he moved among these animals with no weapon and apparently no sense of fear.

Like Thoreau and Marsh before him, Muir argued that everything was interconnected and thus each element was equally essential for the integrity of the whole. Just as Darwin had shown, in a light vein, how the fate of English agriculture turned on the presence of elderly widows whose cats kept the rat population in check, so Muir contended that predators and even insects made a vital contribution to our well-being. He passionately argued for preserving as much as possible of our natural endowment, especially the western mountain ranges whose breathtaking vistas had so inspired him.

For a time, Muir made common cause with prominent advocates of conservation, most notably Theodore Roosevelt and his secretary of the interior, Gifford Pinchot. With their backing, Muir spearheaded the effort to create Yosemite Park, arguably Muir's most notable and lasting gift to posterity. But it was clear that Muir and Roosevelt loved the out-of-doors in very different ways. While he was an acknowledged and avid naturalist in his own right, Roosevelt's appreciation of nature went well beyond contemplating nature's intricacies and wonders. His engagement with nature was also about conquest and dominance. Understanding aided the effectiveness of the hunter and deepened the sense of awe and accomplishment that came from bagging a trophy-sized animal. As for Pinchot, there can be little gainsaying his love of nature and his desire to save it from thoughtless or wanton exploitation. He devoted his life to creating a framework of public policy and attitude that aimed to protect the environment from degradation and abuse. His goal was to *conserve* nature through prudent, scientifically sound management practices so that the nation could dependably draw upon natural resources without fear of depletion.

Though probably no more astute a naturalist than Roosevelt or Pinchot, Muir was, by contrast, committed to *preservation*. He was in love with nature unspoiled. He went afield with little more than the clothes on his back. Roosevelt went out "loaded for bear," literally and figuratively. Muir lived off the land in ways that scarcely left a trace. Roosevelt cut a wide swath. Both were at home in the wilds, in touch with their senses, both were exhilarated and challenged by the majesty and untamed beauty of the wilds. But their respective senses of nature and what it meant to be a part of nature were worlds apart.

Muir, like Thoreau before him, saw nature as fundamentally benign. Of course there was predation—without death there could be no life. Precisely because death was not gratuitous in nature, because death gave sustenance to the living, undisturbed nature had about it a calm, almost beatific aura. Submitting to nature on its own terms, as Muir endeavored to do, thus would have a calming and restorative influence. This is largely why Muir argued so passionately for the need to set aside as much area as possible for parks. Not only would parks preserve our natural heritage and provide habitat for flora and fauna that would otherwise be driven to extinction by the advance of settlements, parks were also places where humans could restore themselves. Much as Walden Pond helped Thoreau to gain perspective and self-knowledge, Muir thought the parks could do the same for all of us.*

Teddy Roosevelt also found nature revitalizing, but he found restoration in the challenge that nature offered. He saw nature as a marvelously complex and stimulating puzzle. Part of this puzzle involved devising ways of conserving nature while using it. He was confident that, through careful management and regulation, natural resources could be renewed in perpetuity, thus ensuring future generations access to the same sorts of inspirations and exhilarating experiences he enjoyed.

---

*When the "other transcendentalist," Ralph Waldo Emerson, traveled to California late in life, Muir arranged for Emerson to join him in a camp in the fabled redwoods north of San Francisco. Muir was certain that the majestic trees would be a fitting backdrop for conversation with the philosopher who believed that humans and nature equally shared in a cosmic soul. Imagine Muir's chagrin when Emerson made plain, after a brief encounter with the forest and almost no conversation, that he preferred the hotel to the woods, and abruptly departed.

Conservationists like Roosevelt and preservationists like Muir could agree on the need to set aside reserves such as Yosemite National Park. But there the agreement ended. From the point of view of conservationists, preservationists stood in the way of progress and growth. Mainstream America was too eager for the easy prosperity that our abundant natural resources offered to take seriously the admonition to leave vast stretches of land alone. By early in the twentieth century, the unsettled lands beckoned to rapidly rising numbers of Americans intent on partaking in some of the same experiences that Muir and Roosevelt made so appealing through their writings.* The railroad companies strenuously promoted the great western parks in order to stimulate passenger traffic. Camping grew in popularity as urbanites sought respite from the heat, foul air, and congestion of the city. And as the automobile became affordable, thanks largely to Henry Ford, more people got into the country on weekends and vacations. In short, the preservationists lost the early battles before they even had a chance to rigorously articulate their position. Indeed, it would not take many decades before the nation's parks would begin to deteriorate under the sheer pressure of numbers. The conflict between preservationists and conservationists clearly made it harder than it might otherwise have been for the nation to formulate a broadly shared solicitude for the environment.

Aldo Leopold, without question the single most important figure in modern environmentalism, struggled to find ways of charting the balance between preservation and conservation. Fresh out of forestry school, Leopold joined the then-infant U.S. Forest Service in 1909 and was assigned to manage the national forests in Arizona and New Mexico. One of the tasks he oversaw was the elimination of wolves from the

*Though Thoreau might well impress us as a more "exotic" figure than Muir or Roosevelt, at the time that he wrote of nature and simplicity, most Americans were in fact living at least as close to nature as he was. They may have understood that closeness differently, but the idea of living in a rough-hewn cabin and relying on foraging to supplement produce grown in the garden could scarcely have seemed remarkable to most Americans in 1850. By 1900, though, taking pack horses into the backcountry to hunt elk and bear or heading out for the Tuolumne Meadow with no provisions and only the clothes on one's back had to strike most people as exotic. For an excellent discussion of the shift in American attitudes toward nature and the wilderness at the turn of the century, see Peter J. Schmitt, *Back to Nature: The Arcadian Myth in Urban America*, Baltimore, MD: The Johns Hopkins University Press (1969).

forests and range lands of the Southwest. Later, in his famous essay "Thinking Like a Mountain" (1944), he reflected back on a moment in that campaign. "I was young then and full of trigger-itch; I thought that because fewer wolves meant more deer, that no wolves would mean hunters' paradise."[7]

The need to recast our alternatives and rethink things struck him while gazing into the dying embers of a she wolf's eyes that he had just shot. He arrived at the wolf's side and "watch[ed] a fierce green fire dying in her eyes." The embers in the wolf's eye kindled a fire in Leopold. Over time, he came to realize that wolves, like everything else, played a vital role in maintaining the integrity of an ecosystem. Ranchers had implored the Forest Service to kill the wolves so their sheep and cattle would be safe. But when the wolves were eliminated, the deer population rose, the rabbit population exploded, and the end result was that the domestic animals faced stiffened competition for food. Wolves and coyotes, as it turned out, dined only occasionally on lamb chops or filet mignon. Most of the time, they made do with rabbits, mice, and deer. Kill the predators, Leopold deduced, and you indirectly destroy the range for every creature, domestic and wild.

Leopold's appreciation for the interrelatedness of all life forms was, in itself, far from new. As we have seen, the idea of nature as an intricate web of reciprocities and interdependencies can be traced back at least to Thoreau and Marsh. What Leopold did was begin to transform the loose amalgam of ancient myth, modern sentimentality, and contemporary empirical science into a coherent, empirically based ethic—a statement about how humans ought to regard nature and define their role in the order of things. Central to his emerging stance was a sober acknowledgement. In a 1927 letter to the head of Glacier National Park, Leopold wrote:

> I must confess that it seems to me academic to talk about maintaining the balance of nature. The balance of nature in any strict sense has been upset long ago, and there is no such thing to maintain. The only option we have is to create a new balance objectively determined upon for each area in accordance with the intended use of that area.[8]

In elaborating what he came to call the "land ethic," Leopold rejected the notion that humans can simply stand aside and leave nature to her own devices. We are, after all, part of nature too. We can no more stand aside than can the wolf or the goose. Each of us acts and interacts with the others—there is no possibility of being innocuous. This doesn't mean that we should be noxious. Being ineluctably part of nature doesn't give us free reign to do whatever we want. Rather, it obliges us to *think about what we do in the context of the system as a whole.* Unlike the wolf or the goose, we have choices. We should choose, Leopold reasoned, those actions that do not diminish diversity in the biotic community. We should avoid unnecessary stresses on the environment, and we should allow nature plenty of room to recover, restore, and replenish. We should behave as though there were a tomorrow. We should take the long view or, as he put it, "think like a mountain."

Part of this long view entailed acceptance of the need to manage and manipulate nature. Leopold, without pleasure but also without apology, insisted that it was futile to be mesmerized by dreams of pristine landscapes. Humans had been around too long and had far too great an effect on nature to permit a return to some original state, whatever that might have been. Since we cannot extricate ourselves from nature, we are, he argued, obliged to bring the full measure of our rational faculties to bear in determining what our role in nature should be.

What makes Leopold so special, so seminal a figure, is the artful way his perspective blends genuinely antagonistic points of view. He is at once a scientific manager, an avid hunter, a believer in nature's majesty, and an advocate for what might be thought of as "radical humility." In most hands, such a mixture would become mush. Leopold molded the disparate elements into a rigorous and compelling perspective that has been a benchmark for all serious environmental thinking from his day to the present.

This is not to say that the contradictions Leopold wrestled with, and at least provisionally tamed, have gone away. Were this so, Leopold's name would now be a household word. In part because Leopold wrote little, and in part because much of what he wrote was either highly technical or densely compressed, his synthesis did not quell continued debate along traditional lines. As a result, environmental management efforts are often misguided attempts to recapture a pristine state or thinly veiled covers for commercial exploitation. Critics of management, who have plenty of mistakes to point to, for their part are all too frequently drawn toward a "hands-off" position, imagining that some-how things will right themselves if only we get out of the way and stop trying to manipulate things. Both proponents and opponents of man-agement use Leopold to buttress their claims, each side conveniently forgetting the synthesis that Leopold was pressing for.*

It may well be that such a synthesis is illusory. There will certainly be a constant if not growing need for experts with the knowledge (and/or hubris) required to manage resources, clean up past excesses, and de-sign new technologies to meet future needs, whatever they may be. It is equally certain that these efforts to manage and regulate nature will result in disasters, even colossal ones. These "mistakes" will provide more than ample ammunition to those who reject management on the grounds that we cannot possibly improve upon nature. Whether Leo-pold's goal turns out to be illusory or not, he helped found the Wilder-ness Society in hopes that the little that did remain of our wilderness legacy might be preserved, both for the spiritual values inherent in wilderness and for the benchmarks such areas can provide for deter-mining how we might best manage the rest of creation that we have already altered irrevocably. We cannot recreate genuine wilderness, though we might reclaim aspects of wildness.

The Quabbin as Battleground

On a very small scale, though one in which the lives and livelihoods of several million people hang in the balance, the men and women re-sponsible for maintaining a safe and abundant water supply for Boston labor daily with the burden of managing the environment, of "playing God" with the Quabbin. When, in the late 1980s, they could no longer ignore the possibility that the forest surrounding the reservoir, upon which they depended to stabilize and purify the flow of water into the

---

*For a spellbinding account of how quite contradictory policies have been justified in Leopold's name, see Alston Chase's (1987) report on the disputes that have characterized the management of wildlife, particularly the elk, of Yellowstone National Park. The problem in Yellowstone is, on a much grander and more complex scale, not unlike the problem the MDC faces with the Quabbin. It is ironic that one of Leopold's sons, Starker, himself a professional employee of the Park Service, was a pivotal actor in framing what clearly has been a disastrous response to the growth of the elk herd in the park.

impoundment, was being jeopardized by an inordinately large deer herd, they stepped into the vortex of conflicting values, definitions, and attitudes that characterize our sense of nature.

The immediate problem was simple enough. Inadvertently, the Metropolitan District Commission had created ideal habitat for deer when they prepared the land for the reservoir and declared it off limits to hunters, the only major predator of deer left after the wolves and mountain lions had been driven out. Fluctuating water levels have helped to provide ample food for deer, as have a succession of mild winters. In addition, the management strategies the MDC has pursued have helped to maintain the quality of the deer habitat. Nature and the MDC have ensured that the deer would be superabundant. This has meant, over the long haul, that forest regeneration, at least regeneration of the species of trees the deer consider delectable, would decline.

As we have seen, the low regeneration, particularly of oaks, was noted in the first forest inventory completed in 1963. But there were more urgent things calling for attention back then and the matter was put on hold. Once concern over the adequacy of water supplies abated, however, the problem of regeneration became more salient. It quickly became clear to those surveying the forest that regeneration was approaching nil over large stretches of the reservation. In a nice piece of professional understatement, an MDC report, "Deer Browse Impacts on MDC Quabbin Watershed Lands: Answers to Commonly Asked Questions" (25 October, 1989), declared:

> Even in its 1972 Quabbin Forest Management Plan, MDC noted that in order to reproduce the present forest, significant forest regeneration should be developed within 15–20 years. Since that time, forest decline has occurred at rates much greater than anticipated. In addition, it has become apparent that the establishment of regeneration even without deer impacts would require many years and intensive site preparation efforts in some areas due to the herbaceous layer [i.e., ferns] which has become firmly established in heavy deer browse areas. (4)

With each passing year, the MDC feared, the problem would grow more urgent, not only because the ferns get more entrenched but also because the forest as a whole becomes more mature. If little or no regeneration occurs, the aging forest grows more vulnerable to storms, disease, and pests. The inevitable damage one or more of these forces would cause to the mature trees would have an exaggerated impact because no young trees would be there to take their place. The MDC examined the relative merits of the sort of forest they had compared to the one the deer had them headed for, and they discovered that a deer-managed watershed was decidedly inferior as regards the all-important issues of water quantity and quality.

The pace of studies picked up, and by 1989 the matter was settled: the deer herd would have to be reduced in size. Every conceivable alternative had been looked into or, where feasible, experimented with. Tubes had been placed around young saplings to shield them from the deer. Small areas had been enclosed with electrified fencing to keep the deer away. In each of these cases, new growth flourished. The only difficulty with these solutions was a practical one. On the scale that would be required to address the problem, the costs of tubing or fencing would be prohibitive, and the manpower necessary to check and repair the tubes or fences would tax the personnel of the MDC to the breaking point. Other possibilities considered included birth-control implants, a logistical nightmare given the huge proportion of females that would have to be treated in order to achieve an appreciable decline in birth rates, and relocation of deer off the reservation. The latter option was rejected because it too posed daunting logistical problems, and also because studies had shown that relocated deer experience very high mortality rates, resulting either from the stress of capture and transport, or from the stress of being torn out of the particularities of locale. (Deer are highly sedentary, rarely moving much beyond a half-mile or so radius of the area of their birth.)

Unable to interfere with the deer's access to new growth on the scale needed, and unable to control deer by removal or birth control, the MDC was left with one option: increase deer mortality. Throughout the 1980s, agency personnel had hoped for a return of severe winters that would cause heavy mortality through stress and greater predation. Deep snow cover reduces the speed and agility of deer to a much greater extent than it impedes predators like bobcats and coyotes and roaming packs of domestic dogs. Heavy snow cover also limits the deer's access to acorns and roots, making the young and the very old vulnerable. The snows did not come. On the contrary, winters grew milder.

The MDC undertook a study of the main predators on the reservation, bobcats and coyotes, in an effort to better determine whether these animals could bring the herd down to acceptable levels. The results, interesting in their own terms, were far from encouraging. Though highly efficient at their business, bobcats and coyotes were no match for the fecundity of deer. The MDC even briefly considered introducing larger and presumably more ravenous predators like wolves and mountain lions. The problem with these larger predators is their range. Coyotes and bobcats, like deer, are homebodies. By contrast, mountain lions and wolves need room to move and, big as it is, the reservation would be far too small to contain these animals. MDC officials smiled wanly at the prospect of dealing with someone whose house pet or livestock was lost to a wolf or mountain lion that they had introduced.

As the options for increasing mortality fell by the wayside one by one, it became clear that there would be no alternative but to bring people onto the reservation to kill deer. The only questions were which people, how many, and for how long. Before even beginning to tackle these questions, however, the MDC realized that it was going to have a tremendous selling job to do to get the public to accept killing deer at the Quabbin. As we have seen, the MDC had been at pains for years to cultivate broad and devoted support for the Quabbin as a sanctuary. If any changes were made, the Quabbin's most avid boosters would have to be brought into the decision-making process.

In late 1989, the MDC began holding workshops to which select members of the public were invited. These were by and large people who had long been identified with environmentalism or who had been active on one or another of the public boards and commissions that provide at least nominal oversight of watershed policies. Individuals who had been especially active in the Friends of Quabbin, a group the MDC had worked closely with for years, were also included in these workshops. Though a handful of insiders had had wind of the MDC's concerns, the workshops were the first public declaration of the deer problem. They provided a forum in which knowledgeable and committed people discussed the nature of the declining forest regeneration, the risks posed by low regeneration rates, and the various actions that the MDC had been considering.

The response from the workshops was only partially encouraging. A number of people were skeptical of the MDC's claims. Even those who accepted the agency's assessment worried about the implications of authorizing any kind of shooting and killing on the reservation. Fishing, though it had been going on for a long time, was still a bitter pill for many of the Quabbin's most ardent supporters. The prospect of letting gunners in to thin the deer herd was repugnant to most attending the workshops. For some, it was an outrage. The Quabbin had become a potent symbol of benign nature, a hallowed place were people could walk among unfearful wild animals under broad canopies of lush forest. That something was amiss in this primeval scene, and worse, that the cure might involve killing on a large scale the very animals that had come to symbolize the tranquillity of the Quabbin's "wilderness," was beyond comprehension for a number of the workshop participants.

The constituency that the MDC had labored so long to assemble began to dissolve into sharply divided camps. Those who sided with the MDC were persuaded that the peril to the watershed and, ultimately, to the water supply was great enough to overwhelm any scruples one might have about killing deer. Those who opposed the MDC's position, as we shall see in detail shortly, were not of one mind. Some rejected the agency's concern for regeneration, claiming that as long as some trees and bushes were growing, the watershed would remain functional. Others accepted the lack of regeneration of dominant species like the oak as a problem but pinned the blame on the MDC's forestry practices. Still others were flatly opposed to shooting and killing. Some people of this latter persuasion were opposed to killing animals no matter what, while others were simply against killing on the Quabbin. They were committed to the Quabbin remaining a peaceful sanctuary.

Initially, the MDC had hoped that it could arrange for SWAT-like teams of sharpshooters to come in, do the dirty work, and leave. This option, too, was dropped for logistical and other reasons that we shall explore in detail in chapter 4. That left but one realistic course: to enlist the Commonwealth's licensed hunters to thin the herd, much as they had been doing every autumn for centuries. Some who were willing to abide the "surgical strikes" of sharpshooters balked at the prospect of allowing ordinary hunters onto the reservation. Many feared that hunters would unite with their fishing buddies and form an even stronger lobby pressing the MDC for more access. Not only would such access ruin the wilderness aspects of the Quabbin, it would also preempt many if not all other, more passive uses of the reservation. When a public hunt became the only viable proposal, opposition to the MDC plan and to the agency's entire management strategy hardened.

A new group, the Quabbin Protective Alliance (QPA), emerged as the umbrella under which the various strands of opposition found articulation. The QPA grew quickly, attracting a broad range of angry and disillusioned former MDC supporters, many of whom had heretofore never engaged in any sort of political, much less protest, activity. With few resources other than passion and zeal, the QPA began to produce a systematic critique of the MDC's past practices and to offer an alternative of its own. Characterizing the MDC's management as heavy-handed, the QPA argued vigorously for a minimalist strategy, the centerpiece of which was an end to logging on the reservation.

The MDC had long experience battling the fishing lobby, but they were clearly not prepared for the hostility and wrath that arose from the ranks of former allies. With very few exceptions, those who had always regarded the MDC as a model of environmental consciousness turned on the agency with the anger only a spurned lover could muster. As the QPA gained adherents and criticism mounted, the MDC had no choice but to make its case in the unfamiliar court of public opinion. Having failed to garner support with workshops, the agency decided to hold a series of public forums across the state in which its position could be aired. MDC officials invited their critics as well as key supporters to participate in the forums. In this way, the interested public would be able to hear all sides and draw its own conclusions. In the meantime, the MDC began discussions with the Division of Fisheries and Wildlife and with key legislators to iron out details of the hunt and to start drafting the legislation needed to proceed.

The Metropolitan District Commission held three meetings during the summer of 1990, two in the western part of the state nearest the

Quabbin and one in Waltham, just outside Boston. One final public meeting was held in May 1991 in Belchertown, the town in which the Quabbin administrative buildings and the principal public access to the reservation are located, at which time the MDC outlined the details of the proposed "deer reduction program" that had been hammered out during the preceding months. Once the issue had been thoroughly vented and opponents had clearly had their day, getting the authorization for the hunt through the legislature and signed by the governor would almost be a foregone conclusion. The real ordeal was enduring the public outcry.

And outcry there was. In the days before the first of the public meetings in 1990 was to be held in Belchertown, local newspapers were filled with letters denouncing the MDC, hunters, and hunting. At that meeting, held in the high school auditorium, those of us attending were greeted by placard bearers denouncing the killing of animals ("Killing for fun is obscene," "Stop the war on wildlife") and the eating of meat ("Meat is murder"). Inside the auditorium, supporters and opponents seemed evenly divided. The program began with a panel made up of critics and supporters summarizing their views of the issue. The panelists included Ray Asselin of the Quabbin Protective Alliance, Jennifer Lewis of the Massachusetts Society for the Prevention of Cruelty to Animals, and Robie Hubley, a staff member of Massachusetts Audubon, all critical of the MDC and the hunt. The proponents were represented by William Healy, a U.S. Forest Service wildlife biologist specializing in the study of deer; Charlie Thompson, a professional forester and spokesperson for the Massachusetts Foresters Association; Tom Berube, president of the Massachusetts Sportsmen's Council; and Elisa Campbell, a prominent local environmental activist and elected town official from neighboring Amherst, who had been one of the people brought into the policy debate in the workshops held the preceding year.

The audience was lively but well behaved. The hunters in the audience would nod vigorously when someone made a point they agreed with. Many of them wore baseball-style caps on which they had pinned their hunting and fishing licenses, so when they nodded, a general flapping could be seen throughout the auditorium. Critics were somewhat less obvious in their partisanship, but murmurs of approval and groans of dismay regularly coursed through the hall. The key moment, replayed in each of the two subsequent public gatherings, was when Elisa Campbell voiced her support for the hunt. She was, she made clear, a lifelong opponent of hunting and in general a staunch advocate for people keeping their hands off nature. And she started out with a bias against the MDC. But she had gone to the workshops, toured the reservation with MDC foresters, and read the voluminous reports, and she could, in conscience, come to only one conclusion: the deer had to be controlled, and the only way to do that, much as it pained her, was to kill deer. The MDC critics groaned at the betrayal by one of their erstwhile allies, and the hunters, surprised at the conversion of an "anti," smiled broadly. Some even clapped.

When the formal remarks were concluded, the microphones were turned over to the audience. One by one, men and women came forward to make their views known. Some were highly articulate and combative. Most were hesitant and torn. Some, like Campbell, hated hunting but feared that the deer were too numerous. Some hunters expressed a deep affection for the Quabbin as a place that was off limits and didn't relish seeing orange-clad hunters there. Back and forth it went as people grappled with their own convictions and hopes for the environment. A time for adjournment had been announced at the outset of the forum, and as the time approached, people began to drift out of the hall, sharing impressions with one another in hushed voices, and offering guesses about what would ultimately be decided. Though I have no way of being sure about this, I left with the sense that few minds had been changed. The meeting was more about catharsis than persuasion.

Subsequent meetings were a bit more boisterous. In predominantly rural Barre, a town adjacent to the Quabbin to the east, a large contingent of hunters staged a noisy walkout to register their dismay that the hunt would not be held that very fall. As they grumbled about "talk and no action" on their way out, it was plain that their interest in the matter was a narrow one. In Waltham, animal rights activists predominated and were much more forceful in making their objections known than they had been in either of the other two meetings, where they had been a decided minority. Still, I was struck throughout by the general decorum of the audiences. After all, feelings ran very deep, and each camp had much at stake in the outcome. The MDC had gone public, and despite the general low regard Bay Staters have for their state agencies, it had weathered the ordeal and shown itself to be responsive and open to criticism, even patient with what must have seemed cheap shots from some of its critics.

It was only at the last of the public meetings, held again in Belchertown, that vitriol and anger really surfaced. By May 1991, the writing was on the wall. The legislature had passed and the governor had signed the bill permitting the MDC to open the Quabbin to hunters. The ranks of the opposition were clearly frustrated. Only one small ray of hope remained for them. An animal rights organization, Citizens to End Animal Suffering and Exploitation (CEASE), made it known that it was preparing legal arguments and would seek an injunction to prevent the MDC from proceeding with the hunt on the grounds that deer hunting on the reservation would imperil the bald eagles, an endangered species, that wintered at the Quabbin.* Several speakers tried to get MDC spokesperson Robert O'Connor, the agency's director of natural resources, to acknowledge that the eagles were being put in harm's way. When he pointed out that the rules governing the hunt would keep hunters well back from the shoreline where the eagles perch, the crowd made derisory remarks about hunters never following the rules.

With the exception of a few hunters who had the temerity to rise in support of the need to control the deer herd, the crowd was decidedly hostile. Speaker after speaker denounced the MDC for its callousness and for running roughshod over the environment. Some speakers were only slightly more charitable, alleging that the MDC had had the hunt forced on them by the hunting lobby. O'Connor listened patiently, rarely letting his emotions show, and, where appropriate, tried to ex-

---

*CEASE filed suit just days before the hunt was scheduled to start, hoping that even if they ultimately lost on the merits of their case, the judge would issue a temporary stay. The judge, after hearing opening arguments and reviewing the evidence that was to be submitted, refused to issue a stay and denied the injunction. CEASE vowed to try again. Indeed, in the summer of 1992, they once more filed suit in an effort to stop the hunt. This time, full proceedings ensued and the court upheld the MDC. CEASE has indicated that it is thinking about an appeal, but as of this writing there has been no further legal maneuvering.

plain yet again why this or that alternative to hunting had been re-jected. One particularly agitated middle-aged man who, so far as I could tell, had attended none of the previous public meetings and certainly had not spoken publicly before, took the floor several times to berate O'Connor and the MDC for permitting the killing of innocent creatures. He urged that the MDC instead lace salt blocks with an unspecified birth-control agent and distribute them throughout the forest. When O'Connor gently suggested that to disperse large quan-tities of birth-control compounds with no means of regulating the dosage or the access of nontarget species, not to mention the possibility of the compounds leeching into the reservoir, would not be responsible watershed or wildlife management, he was assailed for minimizing the harm guns do. Others urged that the matter be put to the voters. Still others, some barely able to control their emotions, refused to believe that deer could ever be a problem.

This was the common thread of the criticisms to which the MDC was subjected: humans are the problem, not peaceable, beautiful animals. The managers of the Quabbin, once thought to be protectors of nature and models of circumspection, were now cast as villains. The Quabbin had come to stand for all that was hopeful in the treatment of our natural resources. Now it was going to become just another killing field. The deer would join the fish and trees as hapless victims of human folly.

The controversy over hunting on the Quabbin reveals deep divisions over how we conceive of nature and our relationship to natural forces. In the chapters that follow, we shall explore these divisions and what they reveal about our attitudes toward and beliefs about nature. While the Quabbin is in many ways unique, and the controversy around the proposal to initiate hunting there more likely to subside than to esca-late, close attention to the nature of the partisanship that arose will, I think, help us gain insight into the problems involved in determining how best to protect and enhance the environment. It is my hope that the chapters that follow will help us understand and maybe even sur-mount the difficulties we face in achieving an environmental ethic that will guide us in our efforts to protect the natural world on which we all depend.

2

Let Nature Be

> For the people of the next millennium, the qualities of nature-honesty, reliability, durability, beauty, even humor-will be necessary landmarks for survival, there for the finding, unless the damage we are doing now proves too great.
> . . . If we jettison our illusions and false assumptions of power and control soon, we *will* leave intact enough of the natural world to care for those who follow. . . .
> DAVID EHRENFELD, *Beginning Again*

THE DESIRE to manipulate nature for our benefit, long taken to be at the core of human nature, is under increasing attack. The daily news regularly informs us of large and small disasters awaiting the planet as a result of our attempts to modify and manipulate nature for our own purposes. Still, the dominant response thus far has been to turn to scientists and engineers for rescue. We may no longer fully believe in the power of science to unlock all the mysteries of the universe, but most of the time, most of us place our bets on science and technology. Whatever the disasters in store for us, we know that the people in lab coats have performed miracle after miracle and our daily lives, if not our imaginations, are inextricably dependent upon these now-taken-for-granted miracles. Whether we turn on a CD player or a hot shower, we are beholden to those who have managed nature and redefined our relationship to natural forces. However, as any parent can attest, being beholden can just as easily produce anger and mistrust as gratitude and admiration.

As the professionals managing the Quabbin edged toward requesting legislation to permit the hunting of deer on the reservation, the relationship they had enjoyed with environmentalists turned sour. Many people who had been staunch supporters of the Metropolitan District Commission grew bitter and accusatory. Individual MDC staffers were singled out as having "sold out" or otherwise betrayed the public trust. Old friendships were strained over the issue, and some were ended. Having long enjoyed broad general support, the MDC now faced organized and articulate opposition from the Quabbin Protective Alliance. In contrast to the more compliant Friends of the Quabbin, the QPA took a decidedly adversarial stance toward the prevailing management policies and practices of the MDC. The agency quickly realized that managing the deer would be uncomplicated compared to the difficulties of managing the growing ranks of indignant citizens. People who had long seen the Quabbin as an island of serenity and harmony, a place where nature was free to run its own benign course, had to be convinced that there was something desperately wrong on the reservation.

Could Deer Be a Problem?

The signs of overbrowsing by deer at the Quabbin are incontrovertible but not obvious to the casual observer or the uninformed. Nothing is obvious unless you know what you are looking for. In the case of overbrowsing, you have to know something about what a forest looks like when deer are present in more nearly ordinary numbers. To the uninitiated, the forest surrounding the reservoir is a sea of green. With the exception of a few small meadows, some the work of beaver and several others the result of clear-cuts of red pine done in the early 1980s, there are trees all over the place. It may literally be the case that the abundance of trees obscures the forest.

Under normal circumstances, a forest like the Quabbin's would have a profusion of "sticks": small saplings, tangles of vines, and thick patches of wild raspberry and other early successional plants woven in and out amongst the more mature stands of trees. In the summer especially, a visitor to the forest would be confronted by a nearly impenetrable screen of vegetation that would be as hard to see beyond as it would be unpleasant to move through. To be sure, a welter of paths would wind and crisscross through the woods, marking the comings and goings of rabbits, deer, turkeys, and the animals that prey on them. That's what a typical forest would be like in central Massachusetts some fifty to eighty years or so after it had last been cleared.

Instead, visitors to the Quabbin are greeted by a forest that is anything but an impenetrable screen. Where the topography permits, you can see for hundreds of yards in any direction. Beneath the lofty trees there are no "sticks" and almost no tangles. Rather, there are large expanses of leafy, invitingly soft, green hay-scented ferns. Oceans of ferns. It looks perfectly "natural"—peaceful, undisturbed, a veritable forest primeval. Of course it *is* natural, in the sense that no human planted the ferns. But being "natural" is not the same thing as being "typical" or "healthy" or "desirable," though many of the MDC's critics insist that "natural" does carry these additional meanings.

The first thing the MDC foresters had to do was convince people that what they saw was not what they thought they were seeing. Instead of a forest primeval, they were seeing a forest that was no longer reproducing itself because the deer were eating everything within their reach—everything, that is, that they liked. Since they didn't eat ferns, ferns were everywhere. Deer do like the buds and tender shoots of almost all young hardwoods and softwoods, and they have been munching them, virtually all of them, down to the ground. That is why the woods at the Quabbin are so open and afford such pleasing vistas.

Interested members of the Friends of Quabbin and people drawn from regional conservation and environmental groups were taken on walking tours by the MDC staff to see for themselves what the deer were doing. The MDC, for demonstration purposes, fenced off a few very small areas to make the contrast stark. The difference was indeed dramatic. Inside the fenced area a dense tangle of vegetation thrived. Outside the fence, there was nothing but a carpet of ferns.

By the time the issue of the deer heated up and was brought to the attention of the general public, everyone who had been actively involved in discussions with the MDC staff or who had begun to monitor the agency's reports readily acknowledged that the deer herd at the

Quabbin was far larger than was typical in woodlands outside the Quabbin, and that the herd was having an impact on the forest. But agreement stopped there. At the four gatherings called by the MDC to inform the public and to hear its reaction, some people discounted the seriousness of overbrowsing. One person could not see what all the fuss was about because "there are scads of trees out there." Another speculated that the deer would eventually leave the Quabbin on their own, looking for greener pastures, and thus give the forest a chance to produce a new generation of trees.

In one way or another, all the MDC critics discounted the seriousness of deer browsing and were skeptical of the agency's claim that the deer's impact on the forest placed the reservoir in jeopardy. Many echoed the sentiments of Martin Dodge, a soft-spoken retired Ph.D. who, with his wife, has devoted his time and savings to restoring and preserving "natural areas." "Is there a regeneration problem?" he mused.

> Well . . . there's a *perceived* regeneration problem. Clearly there aren't as many trees coming up as perhaps there would be but [it is] only the commercial type trees that aren't coming up as much.

The problem from Dodge's point of view was that the MDC was overly interested in the harvest and sale of oak logs. The other critics, to a person, shared this view. For example, Joan Williams, who, with her partner, Jack, lives across the highway from the Quabbin, walks almost daily in the reservation, and is active in the Quabbin Protective Alliance, flatly rejected the notion that there was a deer problem.

> In my opinion it's not a deer problem, it's a people problem. . . . There are many deer over there, we see them all the time. And for sure there's many deer. And I suppose if you want oak regeneration and that's your main focus, then you have a deer problem. But I don't really care if oaks grow over there or maples or white pine or spruce.

Things are growing on the reservation, the critics said. They may not be the kinds of plants and trees the MDC foresters would like to see, but so what? Why, Dodge and the other critics asked, should oaks be the goal? Most agreed that the answer was revenues from forestry. It was not simply that the MDC foresters were driven by dollars, though.

Many critics reasoned that foresters are trained to prefer certain kinds of trees, a point we shall return to in a different context later. Lorraine Near, a prominent and outspoken animal rights activist, stated the matter bluntly. I had arrived early for our interview and was waiting in her windowless office that was also serving as a shelter for several cats. The air conditioner was laboring with limited success against the heat of the day but was completely failing to clear the air of the heavy odor of cats. After she greeted me, she muttered "I can't stand these damned cats in here." Having established that her defense of animals was not without limit, she proceeded to explain her sense of what had been going on at the Quabbin.

> I am no wildlife biologist or watershed management person, but I've learned enough in the last couple of years at the [MDC-sponsored] workshops to have a really dubious feeling about the existence of a problem. . . . It seems as though the justifications are being made for the hunt because they want to continue their logging practices. They want to re-create the forest out there so they have small, medium, and large trees so they can continue to cut—a sustained yield thing.

Each of the nineteen critics I interviewed zeroed in on the MDC's forestry practices as the source of the problem. The common denominator in their criticism was the allegation that the MDC, in one way or another, had been producing prime deer habitat through its forestry practices. The critics took particular glee in noting that when the MDC was trying to increase water yield, it had regarded the deer as beneficial. Once anxiety over water quantity eased, the agency seemed to its critics to be hell-bent on logging. With deer no longer needed for "pruning," they were regarded as inimical to the long-term stability of logging operations.

Some of the critics had a conspiratorial view of the MDC. Liz Simpson and Jim Ventura, both intensely committed to living in harmony with the land and with animals, had followed the deer controversy almost from the start. A former policeman, Jim now believes in nonviolence. Liz, who now works with children, has, among other degrees, one in agriculture, which she characterized as training in "violent agriculture." In their view, the MDC created that problem over the years with their clear-cutting. . . . The MDC's full of wildlife biologists and state agency biologists and they're the ones that usually promote and support hunting throughout the state. (Jim)

I think if they really wanted to cure the deer problem they'd have to cure their woodlot management problem. Because that's where it [the deer population] stems from, because deer like to feed on these nice small little seedlings that come up whenever they do their logging. . . . They have a hidden agenda that is to allow logging and eventually make that into a recreation area. (Liz)

Jim's suspicions flared publicly in 1990 when he challenged MDC official Robert O'Connor at the last of the three public hearings held on the proposed deer hunt. In an earlier meeting that Jim had attended, O'Connor had referred to a South Carolina watershed that had suffered extensive damage in a recent hurricane. The forest there had been a largely homogeneous one, both in terms of species and age classes of trees, and when the storm came, it leveled everything. Water quality suffered, O'Connor averred, and the watershed was jeopardized. Between the two meetings, Jim had called down to South Carolina and got, he said, a very different story. For him, this was one more piece of evidence showing that the MDC was simply trying to create a sense of crisis, of imminent peril to the water supply, to justify their "hidden agenda."*

Others were more charitable toward the MDC staff, using adjectives like "sincere," "honest," "dedicated" to characterize particular individuals. But they were not forgiving about the forestry practices. Laura Prevost is another active member of the Quabbin Protective Alliance, who, with her husband, Rick, and their two kids, regularly visits the reservation where she pursues amateur naturalist studies and Rick does nature photography. Laura denounced the MDC's forestry in no uncertain terms.

---

*The facts, so far as I can determine them, are both more complicated and less interesting. O'Connor had consulted his counterparts in South Carolina, the people actually in charge of watershed management and monitoring water quality. Ventura, as it turns out, had spoken with a public relations officer with the U.S. Forest Service who was sanguine about the forest recovering but had nothing to say about water quality.

It really makes me mad when I see just how much they destroyed things over there . . . and they've created the situation and they're just going to make it worse. . . . Sure there's a deer problem there . . . not like Crane's Beach, they're not starving to death, they're not wandering into people's yards looking for food. But there's an overabundance of deer and they are overbrowsing the area. But the reason they are is because they're [the MDC] creating so much food for the deer that the deer are able to eat it and they're able to reproduce. . . . If they stop logging, the deer will stop reproducing at [so high] a rate. But they don't want to stop the forestry practices.

Laura's view of the matter revealed a tension within the ranks of the critics. As we have seen, some of the critics were quite skeptical about the deer being a problem. For Martin Dodge, the deer were one among many natural forces at work shaping the flora and fauna on the reservation and he was content to let things take their course. For people like Lorraine, Liz and Jim, and Joan and Jack, who are strongly committed to nonviolence and animal rights, deer were both harmless and blameless. For critics like Laura, by contrast, deer may have been blameless but they were by no means harmless. The damage that they were doing to the Quabbin forest was just a symptom of a far deeper problem—the MDC's forestry.

Indeed, for critics like Laura, a quick and focused reduction of the deer herd in order to get tree regeneration going again would be grudgingly acceptable if the reduction were accompanied by an end to logging. Ray Asselin, who represented the QPA on the panels the MDC put together for the public meetings, was adamant on this point.

Our [the QPA] official position is that if the MDC were to say "okay, we're going to stop managing this forest," then we have no problem with them electing a short-term hunt, maybe five years or so, to get the herd down initially. *But only if they leave that forest alone!*

Had the MDC taken this course, the opposition would have narrowed to those who were solely interested in protecting the deer or who were committed to the idea of the Quabbin as sanctuary, a sentiment we will explore later. But the MDC was not about to stop harvesting trees. As a result, the ranks of the opposition swelled and the tension between the animal protectors and the tree protectors remained almost completely latent. For the critics, the MDC, not the deer, was the problem.

Deer, Trees and Water: the connection challenged

As far as the critics were concerned, the Metropolitan District Commission had begun to create prime deer habitat when it first cleared the valley of trees to make way for the water. While the reservoir was slowly filling, the land not yet flooded burst forth with new growth that attracted area deer. With hunting forbidden and most predators gone, the deer thrived even as the water levels rose. Subsequent forestry practices certainly did nothing to discomfit the deer. Critics who indicted the forest managers for creating the large deer herd also challenged the MDC's contention that there was a direct link between the damage the deer were doing and the security of the reservoir. They rejected as unproven MDC fears that an even-aged forest, such as was being produced as a result of the deer impact on tree regeneration, was less reliable protection for the water supply than an uneven-aged forest. The critics alleged that the MDC management goal of creating an uneven-aged forest had less to do with water than with justifying its logging operations.

Many of the critics were also skeptical about the MDC's assertion of a relationship between a particular type of forest, deer, and water. Warren Edwards, a university-based scientist and a member of the Friends of Quabbin and several citizen advisory boards dealing with water and water quality, had attended meetings, read reports, and gone on tours of the forest. When I asked him if he shared the MDC's sense of urgency, he replied, "No, I frankly don't." He went on to explain why.

> I think their [the MDC's] sense of urgency comes from what they see as a lack of regeneration and they think that if that continues there are parts of the Quabbin that will convert from forest to herbaceous growth. There are others who say that succession is not necessarily the only route, there may be other kinds of cover, other types of forest that would regenerate. . . .After all, the forest came back after the initial clear-

cutting and deer were numerous then. . . . It all happened very fast. . . and the place didn't disappear because it was clear cut. So I guess I don't share the sense of urgency or feel that we really need to take dramatic action.

Chris Leahy, who served as a spokesperson for Massachusetts Audubon on one of the panels, was more pointed on this matter.

And so I think that there probably would be a difference, there would be more nutrient input from an unmanaged old-growth forest, but I very seriously question [whether] the difference between that input and the quality of the water that you would get from that kind of system would be significantly worse than the more managed forest system.

Such doubts about the conclusions the MDC had come to were broadly shared among the critics. Interestingly, the doubts arose from two opposite responses to science. The critics included several scientists and trained ecologists and wildlife biologists. For the most part, they felt that the MDC had not done its homework—it either had not done enough studies or had not done the right kind of studies. In the view of these professionals, the agency's sense of urgency was not well founded, and the need for a hunt had simply not been established. Jennifer Lewis, staff wildlife biologist for the Massachusetts Society for Prevention of Cruelty to Animals (MSPCA), drew this conclusion, but surprised me when she said that there was a regeneration problem. Here's how she explained her position.

You were asking if I think there's an overpopulation [of deer] problem. I think there's a regeneration problem. I mean there's clearly a regeneration problem on parts of the reservation, especially the lower part of Prescott Peninsula. [But] the standard they're using for judging whether regeneration is adequate is a standard that's used in commercial harvesting operations. And that standard may be too high for their stated objective out there, which is maintenance of the forest for watershed purposes. So maybe they don't need two thousand stems per acre, you know. Maybe they only need a thousand. Maybe they only need five hundred. . . . Whether there's a population problem or not, I think is an unanswered question. And it's possible there may be. I don't know. But I don't think they've asked the right questions or done the right research.

Massachusetts Audubon's Hubley insisted in his panel presentations that the MDC had not done anything like the kinds of studies that would be required to conclude scientifically that there was a need to kill deer. He urged the MDC to do basic research and then develop a "comprehensive ecosystem management plan" before embarking on any new management initiatives, especially ones that would increase public access to the reservation. Mass Audubon staff were clearly worried lest the Quabbin slip steadily away as a potential natural area to become just another "recreational asset."

Most of the critics, while sharing these concerns, were far less concerned about the adequacy of the research the MDC relied upon than they were suspicious of science in general. For many, this questioning of science stemmed from their attraction to the craft of the naturalist. Responding to the reverentially close observation of flora and fauna in their natural settings and marveling at the subtle ways everything seemed to fit together, these people had a visceral distrust of the scientist who is forever abstracting, reducing, and fragmenting things. In the process science loses sight of what is crucial: the way the whole functions.

Bill Granby is a self-taught naturalist and a devotee of eastern old-growth forest. He had not involved himself directly in the Quabbin controversy because, as he made clear, the forest at the Quabbin had been cut over so many times that it held no interest for him. I decided to interview Bill because several of the critics that I had spoken with indicated that he had been pivotal in helping them shape their own understanding of what was at stake in the Quabbin controversy. As much as any one, Bill was responsible for making old-growth forest the rallying cry of the MDC critics. Retired from the Armed Forces, Bill spends as much time as he can seeking out stands of old growth, leading tours, and speaking to groups about the importance of old growth. He is lively and intense, his slight southern accent at variance with his rapid-fire conversational style. He had little good to say for the scientists he has encountered, both as a student years ago and now as a self-taught and self-declared forest ecologist.

> Once upon a time science to me was God. Science had all the answers. Now, in a strange kind of way, from the study of old growth, I see that

there are only questions, there are no answers. So science can't be that godly. I finally ran out of patience with the next particle. . . that was a product of a mathematical equation. Bullshit. I backed away from science as everything. That leaves a lot of holes and how I deal with it is a lifelong occupation.

About the science of forestry in particular, Bill said:

I've seen the many mistakes they've made. Sure, they have a broad base of knowledge. Sure they can understand a lot about plant genetics. Sure they can grow taller individual trees. But they don't really know what the forest ecologists know, because they don't study that.

Martin Dodge echoed Bill's doubts about science, though Dodge spoke more as an insider, having worked for years in industry as a scientist.

I don't think they [the MDC foresters] know enough about the dynamics of the whole ecosystem to really know what they're doing. All they've studied is how to improve a stand of trees, or how to make sure that the trees they want will grow better. . . . Well, I'm sure it's [science] part of the problem. I liked science better when it was people just out trying to find out how things ticked, rather than having a goal that benefitted man and not paying too much attention to the side effects.

Science, it seems, is devoted to breaking things down into ever smaller components, the better to manipulate this or that. For many of the critics, the task was to undo the results of the specializations of science, in effect to put Humpty Dumpty back together again. Because it tended to lose sight of the whole, of the myriad interconnections and delicate balances that constitute nature, these critics saw science as a danger. Dorothy Reading, a young woman who spends a great deal of time hiking in the Quabbin in order to better appreciate the interconnectedness of life, was outspoken on this. "I think that the further scientists go with trying to control these natural cycles, the worse catastrophe there is going to be in the end."

Many of the MDC's critics viewed nature as sacred and elegantly intricate. Places like the Quabbin present the possibility of recapturing a lost purity, of restoring at least a portion of our otherwise abused and neglected landscape to its natural state. In the process, we would, according to some, be taking a small step toward healing ourselves as well. In other words, while welcoming the criticisms that professionals like Lewis and Hubley leveled at the MDC, most of the critics I interviewed would rather see less research and more heartfelt emotional engagement with nature.

No one exemplified this attitude better than Peter Gomes. Bill Granby helped many of the critics appreciate the virtues of old-growth forest but Peter was unquestionably the spiritual force behind the opposition to the proposed hunt. People spoke of Peter in admiring terms, sometimes bordering on the worshipful. (This admiration was not universal. Some supporters of the MDC—and some, but by no means all, of the MDC staff—thought he was, as one fellow put it, "a nut.") Peter has become renowned for his animated lectures on natural history as well as for his acute observational skills. In his talks, he mixes philosophy, evolutionary biology, and folklore into an engaging lesson in humility before the altar of nature. He and his wife live on a remote stretch of river in central Massachusetts, connected to the wider world only by a telephone and an aging Subaru wagon.*

Fully at home in the woods, he began life in the ethnic enclaves of southeastern Massachusetts. Disinterested in school, he drifted into teenaged rebellion that blossomed into an early adulthood of hard living—the fraternal order of sex, drugs, and rock 'n roll. A brush with the law and the prospect of prison sent him to the woods to reclaim his life. So far as I could tell, the reclamation could scarcely have been more complete. Peter is, in a word, charismatic. His dark eyes fix his listener and his words flow passionately, especially when he talks of the need to heal the earth and with it, ourselves. He is intimately familiar with the Quabbin and has even been commissioned from time to time to use his observational and woodcraft skills to assist the MDC in conducting censuses of wildlife, particularly predators. Our problem, in Peter's view of things, is that we've lost our capacity to feel a part of nature. We've let thought govern feeling; it should be the reverse.

---

*There is another connection as well. The river is a draw for area fishermen. Though beautiful, it is contaminated by PCBs. At the roadhead thirty yards from Peter's home, a sign warns fishermen not to eat their catch. The irony is not lost on Peter, who gets angry at the mere thought of the effects the contamination is having on the osprey and eagles as well as the raccoons and other fish and crustacean eaters that inhabit the area. "They can't read the warning signs," he said ruefully.

Well, we have the ability to think. Our thinking, to me, is our major problem. Not only has it brought us plenty, and made us the most powerful predator, but it's also going to be our demise. I really believe it. We're so disconnected from who we are, from the reality of our being. We're separate from our thought. The thinker of the thought is separate from the feeler of the feeling. Thought has become absolute and it's separated itself and thinks it's in control. . . .

The Critique of Management

The Metropolitan District Commission failed to convince its critics that deer threatened the water supply. Instead, the critics saw the attempt to manage the deer as an extension of the desire to manage the forest, to sculpt it to suit some abstract, humanly contrived notion of an "ideal forest." They contended that the deer posed at most a very remote threat to the reservoir, and that in fact an *unmanaged* forest would do as good and perhaps an even better job of protecting the water supply.

This desire to see an end to the manipulation of the environment at the Quabbin linked virtually all critics of the MDC and lent coherence to their critique of the proposed deer hunt.* The critique of environmental management stemmed in some measure from the more general skepticism about science that we have just explored. In larger measure, though, the rejection of management seemed to originate in a diffuse sense of dismay with modernity itself. Three themes reverberated through the remarks and observations of the MDC's critics: we do not know enough to manage wisely or well; we are too shortsighted and too preoccupied with selfish, immediate goals; and, finally, managers have become entrenched and self-serving, obsessively committed to tamper-

*For some of the critics, this was a deeply held belief that was generalized to the environment as a whole. Most, though they had some emotional sympathy for the robust assertion of nonintervention, limited their noninterventionism to the Quabbin. As we will see shortly, almost everyone I interviewed agreed that *some* areas ought to be left alone in as pristine a condition as possible. The question that divided folks was whether the Quabbin, given that it was, above all else, a vital source of water for millions of state residents, was an appropriate area for a completely hands-off policy.

ing with nature. In short, we humans are ignorant, blind, and addicted to meddling.

The disenchantment ran deep, though there were important nuances and modulations. As we have noted, several of the MDC's critics were trained in the fields of forestry and biology and were professionals in the general field of environmental policy. The critique of prevailing management policies by people like Jennifer Lewis and Chris Leahy emphasized the ignorance of the MDC staff and the agency's unwillingness to do the kinds of thorough studies that should be done. But they were clear that management is not inherently bad—good, sound management can be had if we are willing to put in the time and effort and do not allow ourselves to be mesmerized by quick fixes and shortsighted expedients. Here is how Jennifer Lewis characterized the matter:

> You know, they have studied the trees to death out there. They have done an incredible job [on the trees], an unbelievable job. All this professional forestry information . . . is great, it's wonderful. They should do the same with the deer. They have extremely rudimentary data on the deer. I mean, they have a very poor estimate of how many there are and a very poor estimate of what the growth rate is. They don't know what's going on—they don't know what immigration is, they don't know what emigration is. . . . I think it's their failure to put resources into looking at the deer population because it hasn't been important until [the deer] started causing a problem. It's their failure to look at the impact of their past management practices on their deer population.

For Lewis and a few others, the problem was partial ignorance, which was, in principle, correctable. Most of the critics, though, were inclined to see ignorance as a permanent condition—we will never know enough to be trusted with such a precious and delicate thing as nature. Martin Dodge expressed this view succinctly:

> I don't think they [the MDC] know enough about the dynamics of the whole ecosystem to really know what they are doing. . . . It seems to me that just about every time we try [to perfect upon nature] something goes wrong, something that we haven't anticipated.

In effect, we are shooting in the dark. Our knowledge is always partial and fragmentary. Even our most well-intentioned efforts are likely to return to haunt us. Laura Prevost, an avid newcomer to nature study, thought that "nature did fine before we came along. What makes us so self-righteous to think that we can play God?" In one way or another, almost every one of the critics echoed these sentiments.

Joan Williams, ordinarily soft-spoken, grew animated and agitated when she began to talk about efforts to manage. When I asked her if she thought nature was too unpredictable to permit us to forego managing things like water and forests, she replied emphatically.

> My response to that, really, is *bullshit*. . . . What about the unpredictability of human beings? Never mind nature. I mean look at what we've done, look at what we've done! We've done much worse than volcanoes and hurricanes and tidal waves and everything else put together. I mean we've done it ourselves.

Jim Ventura, the former policeman, had a similar sense of things:

> Nature knows best. We're always in there trying to interact and create something perfect and it doesn't work. Just causes more problems. When you look at the world, we're the cause of everything that's gone wrong in this world. We did it. It just didn't happen by itself. It was fine before we got here, before we decided we'd have wildlife management, or environment management, or resource management.

The denunciations of hubris were often tempered by the recognition that even activities as seemingly innocuous as gardening are a form of human intervention, a kind of management. Jim and his partner, Liz, distinguished between their organic gardening and the "violent agriculture" that our society has pursued. They have a small plot in a community garden and were proud of how much their forty-by-forty-foot garden could produce. But they were managing, they thought, in a way that was "natural" because it didn't change the fundamental character of the land or soils. As they saw it, they were working hand in hand with nature, not in opposition to it. Joan Williams, whose denunciation of management we just heard, was weeding her garden when I pulled up for our interview. She freely acknowledged her managerial impulses—"Oh yeah, I manage my garden as much as I can." She insisted, though, that hers was management with a "little 'm'," not the "big 'M'" of the MDC.

Scale, of course, does matter. Tending a lawn or cultivating a garden may proceed from the same basic urges that, in other contexts, issue in irrigating vast areas or in clear-cutting a stand of red pine or Norway spruce, but the consequences are as different as night and day.* Still, the possibility that the impulse is the same in both instances cannot be blinked. Perhaps that is why people did not emphasize the differences of scale. Instead, they distinguished their attempts to shape nature to please themselves from environmental or resource management in terms of motives and intentions. The modest scale of gardening went along with a modesty of appetite.

By contrast, managers were seen as the handmaidens of a culture whose appetites knew no bounds. No one held the MDC responsible for the unquenchable appetite of the American consumer, but the critics were inclined to see agency policies as insufficiently resistant to the continual demand for more. Implicitly at least, critics wished that the MDC would take a leadership role in urging greater efforts to preserve rather than exploit nature. Critics reasoned, for example, that if greater efforts were directed at conserving water, the MDC would have no need to increase the intensity of its management activities.

As we have already seen, the critics held that past MDC efforts to milk as much water out of the watershed as possible were a major factor contributing to the continued high deer population on the reservation. Had the MDC not done the clear-cutting of the red pines to increase water yield, the deer would have had less browse and, as a result, lower rates of reproduction and survival. Management, seen in this light, was totally intertwined with the culture of consumption that demands immediate gratification and shows little concern for long-term consequences for resources or the environment. Rather than being regarded as advocates for a more responsible treatment of our environment, resource managers in general and the MDC in particular were indicted as "coconspirators" helping to sustain the illusion that we can, through science and technology, have our cake and eat it too.

---

*Only one of the people I interviewed in fact saw the care of lawns and gardens in the same terms as forestry and wildlife management. The animal rights activist Lorraine Near told me that she could bring herself to mow the lawn or weed flower beds only with great difficulty. The grasses and flower beds are teeming with life that she was loathe to disrupt, much less kill.

Laura Prevost was speaking for most of her fellow critics when she denounced the unbridled appetite for resources.

> I think greed has a lot to do with the motives of a lot of people. You know, like the cutting of trees. It's money. Everything's money. And it's sad how our society has come down to that. Everybody wants more, more, more. And you know, have more money, more money, more money. . . . [We should] go back to the old barter system. I have something you want, you have something I want—let's trade. You know, eliminate a lot of this [consumption].

Greed is all too regularly accompanied by shortsightedness. Lenore James, a young woman whose letters to a local paper caught my eye, brought this out in our interview. Though she and her husband live comfortably in a large, well-appointed house, she clearly understood that this comfort may be contributing to our larger problems.

> Well, at this point I think we're messing things up because we're not looking at the long-term results of what we do. When we invented the air conditioners or refrigerators we thought that was great. Who was to know what kind of havoc that could wreak on our ozone layer? . . . [P]erhaps we should create things not just to better our lives but better the lives of our children and other generations . . . which might mean doing with less.

Resource managers, all the critics agreed, are firmly embedded in the logic of resource exploitation. Whether it was for using commercial standards in judging forest regeneration or for depending upon studies designed to promote the reproduction of game animals, managers were accused of what Thorstein Veblen, the great American social critic, referred to as "trained incapacity." The perspective of managers was, for the critics, too narrow, too technical, too cut off from an appreciation of the whole. Jennifer Lewis, herself trained in a wildlife biology program, spoke at length about the biases of resource managers.

> Well, I think it's a complex situation. And I think it's changing, which makes it even more complex. . . . [F]or many years, wildlife managers tended to come from rural areas and they grew up hunting and fishing and trapping and they didn't see anything wrong with it. . . . And they went into wildlife biology so that they could continue to have that kind

of rural contact. . . spend large amounts in helping wildlife. . . . I think that there is also a bias created by the funding sources [hunting license fees], reinforced by their training, because I was . . . in a very conventional wildlife management program. This was never stated outright, but you know you're there to provide game species to hunters.

Critics like Lewis were sympathetic with their MDC counterparts. Indeed, they were quite willing to praise the agency's devotion to careful logging and its general solicitude for the integrity of the reservation. But however praiseworthy, the MDC was still perceived as too committed to the ethos of intervention: when less might be best, the MDC was inclined to do more.

Other critics were far less appreciative of the MDC's efforts. They questioned the motives of the MDC, especially the forestry personnel. Even when acknowledging the conscientiousness displayed by the foresters, some of the critics condemned the "harvest mentality" they thought characterized the MDC. Martin Dodge saw a kind of cruel irony in the care with which the MDC foresters worked.

For a while they were just kind of caretakers. . . . but then they got their own forestry crew because, you know, there [are] sensitive areas that have to be handled a certain way, and that was all very laudable but suddenly now they've got to support this crew of loggers and foresters. So they've become much more interested in commercial type trees coming up. . . .

Others referred to the expensive equipment the MDC purchased in order to facilitate the highly selective cutting they desired. Ordinary equipment used to drag logs out of the forest creates an enormous amount of collateral damage, damage which the MDC was obviously interested in avoiding. Ray Asselin, the chief spokesperson for the Quabbin Protective Alliance, speculated freely along these lines.

I don't know why that's happened, but [the MDC people] in published articles not too long ago were very protective of the Quabbin. They were quoted as saying we don't need hunters in there to control the deer as long as we encourage the predators and that kind of thing, and they were really very well respected, probably the most respected foresters in the state because of their attitudes toward forest management. You

> know, they wanted to control the loggers so closely that they were not
> well liked by a lot of people because they were so tough. . . . The way I
> understand it, they weren't happy with the quality of work they were
> getting out of those people [private logging contractors]. They couldn't
> supervise them closely enough. So they wanted to get their own crews to
> do the logging . . . and they finally convinced the state to . . . buy
> equipment and hire crews. I don't know if it's coincidental or not, but
> since they got the equipment and crews they've been gung-ho on get-
> ting hunting in there. . . . I think they promised that they could pay for
> the equipment and crews and I suspect that they now see that unless
> they can cut a certain amount of oak every year, they are not going to be
> able to meet that promise.

In other words, the MDC was committed to a management strategy
that, despite its intentions and values, compelled it to get deeper and
deeper into intervention and manipulation, if only to preserve its own
budgets and jobs. This image of getting in deeper and deeper ran
through most of the interviews I had with critics. Many were quite
exasperated. Critics could cite one example after another of how management policy of one period turned out badly, only to be supplanted
by another policy that had equally unhappy results. From this they
concluded that the problem was not just a particular management
policy, not even a particular manager, but management in general. Ray
Asselin again.

> Their management has to get more and more intense to correct all the
> previous mistakes. You know, if they go and harrow the forest floor to
> break up the ferns, what's that going to lead to? Are they going to . . .
> loosen up all that soil, which is going to disturb root systems of every-
> thing that's there . . . ? Just one thing after another is my point.

The critics were in accord on the allegation that the deer herd had
grown so large because the MDC had interfered with the natural processes that would otherwise have kept populations in check. The abundance of deer was the result of misguided management practices, which
would only be perpetuated, or intensified, if the MDC pursued its plan
to regulate the deer by hunting. Enough is enough, the critics in effect
were saying. While there may be little hope of extricating ourselves

completely from the obsession to manage, they insisted that we ought
to be able to hold the line, at least in an area like the Quabbin.

Among the opponents of the hunt were several former residents of
the Swift River Valley who had been displaced to make way for the
impoundment and surrounding watershed. They added their own poi-
gnant note to this view. For them, the Quabbin was a monument to a
way of life now utterly gone but from which lessons remained to be
learned. The tranquillity of the Quabbin stood in sharp juxtaposition
to the hurried and worried pace of life outside the reserve. For these
older people, the only management that seemed appropriate was that
required to keep the area from being abused by thoughtless individuals.
Carlton Wagner had moved with his family from one of the towns
slated to be flooded when he was fifteen and had maintained a lifelong
connection to fellow former residents. After making clear his displea-
sure with the track record of the MDC ("we [humans] have demon-
strated our inability to handle the area"), he went on to observe:

> I feel that it [the reservation] should be left alone. It's one of the few,
> very few areas of that size in this part of the state which, if left alone,
> would develop natural wildlife. . . . We should preserve it and the only
> way to preserve it is to keep the bulk of the people, in particular the
> hunters, out of there.

All in all, the critics despaired over the direction in which they saw
the MDC taking the Quabbin. Land they had grown accustomed to
thinking of as wilderness and sanctuary was being threatened by the
very people who had been its protectors. The critics felt that the MDC
had betrayed them and the public trust. Anxious to forestall additional
and increasingly noxious interventions, they consistently endorsed the
view that the best management was no management.

It would be wrong, though, to conclude that the critics were entirely
of one mind. As should be clear, the proposed hunt posed different
sorts of affronts to different people. The indignation drew everyone
together and provided a common "enemy," but distinctly different
agendas were apparent among the critics. The most modest was that of
the people who had come to love the Quabbin as a sanctuary. For them,
the Quabbin was a living shrine, a hallowed area that served as a

memorial to all that had gone before. They objected, more than anything else, to the *idea of gunfire breaking the peacefulness* of the reservation. As Carlton Wagner put it:

> Yes, gunfire definitely violates the sense of peacefulness there. It's been a
> sanctuary, sort of an island of sanity, and if hunters get in there it will just
> become another recreation area. They forced all the people to move out
> and now they'll let even more back in. That's why I mentioned that the
> first restaurant that goes in, I would like to be the proprietor. [Laughing]
> Maybe I could buy my father's store back.

Former residents like Carlton Wagner believed in progress, and
regarded the Quabbin Reservoir as part of the inevitable march of
civilization. But opening the reservation to hunting was not progress. If
anything, the need for the hunt, about which they remained skeptical,
represented a failure of management to anticipate the consequences of
its own actions.

Opposition of this sort was mild, more a "tut tut" than a condemnation. Essentially, the critics who stressed the tranquillity of the Quabbin would have preferred some unobtrusive intervention if intervention
was indeed needed. They had no objection to the modest logging
operation the MDC was engaging in because they had no principled
objection to viewing nature as a source of vital resources needed to
maintain society. As long as the utilization of resources could be managed discreetly and respectfully, they would be satisfied. But an open
season for deer hunting was, to their way of thinking, neither particularly discreet nor respectful.

In contrast to this mild, though deeply felt, criticism, the opposition
of the animal rights advocates was adamant and unyielding. For the
most dedicated and militant among them, the killing of deer for any
reason was unconscionable. Some among the animal rights constituency could grudgingly accept subsistence hunting for immediate survival needs, especially if the practice was embedded in a traditional
culture. But the worst imaginable violation of their ethical code was the
killing of deer for sport. The salient issue for these critics was pain,
suffering, and death inflicted on innocent individual animals for pleasure. The deer, to use the metaphor of jurisprudence, had committed

no crime, had done no wrong. They were minding their own business. We humans should do the same and let the creatures be.

Our needs or wishes, from this vantage point, are no more compelling or urgent than the needs and wishes of the deer (or any other creature, for that matter). We need to learn how to share the planet with our animal cousins and stop seeing animals as enemies or as objects for exploitation. Animals, from this point of view, have the most fundamental right, the right of liberty, the right to be left alone. This view is not only radical in the sense that it flatly rejects the long-standing distinction between "man" and "beast," upon which most human self-understanding has been premised; it is radical also in the sense that it is radically individualist. The morally relevant unit is the individual animal, not the species, not the ecosystem.*

The insistence of the animal rights activists on the preciousness of each individual animal's life exasperated some within the informal alliance that formed in opposition to the MDC. But many of the opponents of the hunt were willing to accept the notion that nature has a right to be treated with respect, even a right to be left alone under some circumstances. As Laura Prevost's interests in nature deepened, for example, she grew more and more upset about hunting. At first it was an objection to the callousness of shooting something ("that just seemed very cold to me"):

> But now, it's the animal aspect too. . . . I feel we're all God's creatures, and nobody has the right to kill anybody else. . . . I just hate to see anything suffer, you know. . . . There used to be a time I'd flip out if there was a cricket in the house. The other day I'm trying to pick one up in a Dixie cup to bring it outside so it doesn't get killed.

Laura's husband, Rick, has been moving in the same direction.

> I've reached the point where philosophically I feel the animals have a lot more rights than what we've given them. The things I've learned over

---

*Lorraine Near, the most militant of the animal rights advocates I interviewed, made this radical assertion of rights plain when she described the deterioration of her marriage. Her now estranged husband had gone along with the progressive unfurling of her commitment to animal rights, but when she insisted that he stop removing ants from their house (he had already agreed to refrain from putting out ant catchers, instead picking them up and putting them outside), he drew the line and declared he could no longer abide the life style her commitment entailed.

the years, over the past few years from doing these studies having to do with deer and deer hunts and old-growth forest, [have] really made me so much more conscious of the entire ecosystem and the animals and what they all stand for. And you know, I look at it now and tell myself *leave nature alone* because it is beautiful as it is. Let's leave it alone and enjoy it the way it was meant to be.

Thus, the animal rights position was by no means beyond the pale so far as most of the critics were concerned, though, as Rick's mention of "the entire ecosystem" indicates, there was some tension within the ranks. For most of the MDC's critics, the crucial unit of analysis was not the individual animal (or tree), but the *ecological system as a whole.* Individual animals may be expendable if the integrity of the system requires it. As Steve James, whose wife, Lenore, is a vegetarian and very sympathetic to the animal rights position, put it:

I don't have the same ultimate [position as my wife], you know: you can't kill an animal. I . . . have a problem with hunting as a people issue. . . . Some deer really ought to be killed because the herd suffers if there are too many. And the ecology suffers at some point . . . so a humane kill is I guess acceptable. I feel okay to a certain point eating humanely killed meat. . . .

Indeed, the critics most deeply concerned with the Quabbin as a forest ecosystem were, as we saw earlier, willing to endure a quick and efficient reduction of the deer herd *if* the MDC would stop harvesting trees. But if they didn't stop logging, the deer would simply rebound and there would be two harvests in perpetuity: trees and deer. Joan Williams, who is not active in the animal rights movement but who is very sympathetic with it, expressed her feelings this way:

If they really want to cut down the deer herd they ought to just do it and not fart around, you know, having a five- or ten- or fifty-year hunt. And they ought to adjust their forestry management plans.

This reluctant willingness to countenance certain interventions suggests that however jaundiced the opponents of the hunt were toward environmental management, some had at least a small residuum of appreciation for the dilemma faced by those charged with managing.

While many clearly wished to stop all management, others were not prepared to go all the way. Laura Prevost was willing to accept management so long as it was devoted to rectifying past mistakes. Dorothy Reading was willing to accept "minimal management" when, as at the Quabbin, there was an obvious imbalance—when, that is, the intervention was aimed narrowly at restoring balance. Chris Leahy of Mass Audubon did not rule out some form of deer control using lethal means if that proved to be consistent with maintaining the integrity of the whole ecosystem, but he was quick to add that whatever was ultimately decided, less intervention and manipulation were to be preferred to more.

Ray Asselin acknowledged that the deer were a problem and doubtless stood in the way of attaining a healthy forest. He concluded:

> I would tolerate a few years of the hunting, just to get things over with, *if they said they would stop managing the place.* I would tolerate it. But the logging equipment, really, the forestry management issue bothers me more than the hunting issue actually does. . . . I guess the hunting plan was the straw that broke the camel's back for me.

Clearly, such a view would be anathema to someone who believed animals have rights. To my knowledge the critics of the MDC never met as a group to discuss such differences. They had little incentive to do so since they had a very focused agenda and there was sufficient consensus among the factions to foster a unified opposition to the MDC hunt proposal. An enemy of my enemy is my friend, seemed to be the rule.

The different agendas and emphases must be read in a larger context: the critics, to a person, felt that the MDC, not the deer, was the real problem. They were convinced that things would be far better, at the Quabbin and in the world at large, if we would stop intensively managing the environment. Again and again, the men and women I spoke with made plain their desire to see an end to efforts to improve on nature. The following phrases recur in interview after interview: "let nature take care of herself," "things were fine before we started messing around," "nature provides," "we must let nature heal herself," "nature knows best." Carlton Wagner, the former resident quoted earlier, summed up this widely shared belief.

Well I feel nature is a better regulator than man has ever been . . . the deer problem would eventually cure itself much more rapidly than we can do. . . . We've proven our inability to cope with nature. I think nature is a better manager than man will ever be.

Nature's Balancing Act: Old-Growth Forest

The critics of the Metropolitan District Commission were convinced that the Quabbin would be better left alone. The deer and the vegetation would work something out that in the long run would be more stable and durable than anything humans could contrive. The rallying cry of the critics was "old-growth forest." Almost all of the critics I interviewed explicitly mentioned "old-growth forest," often in reverential, even rhapsodic tones. An old-growth forest is, above all else, an undisturbed forest, a forest that is not groomed or managed by humans in any way. It represented, for MDC critics, the complete antithesis of a managed forest of the sort the agency was attempting to maintain around the Quabbin.

But what, exactly, is an old-growth forest? When I first began hearing the phrase I was puzzled. I had associated old growth with the Pacific Northwest, the home of once-vast stands of virgin forests that tower skyward. It is doubtful that the eastern seaboard ever boasted such forests. But even if there were such stands in the East in the remote past, they have long since given way to forest species adapted to repeated disruptions from storms and humans. Most woodlands in the East have been cut at least twice since the English began colonization, and many areas have been logged over three or more times.

Bill Granby was only too happy to inform me of the many forms old growth can take. He confirmed that eastern old growth is not the same as that of the Pacific Northwest—weather, soils, and topography dictate different species, different growth rates, and differences in mortality. He suspected that many people walk right by eastern old growth, unable to recognize it because their image of old growth, like mine, is locked in on scenes from Oregon or Washington State.

Another reason people do not notice old growth in the East is that, with the exception of a large tract in the Smokey Mountains, old growth stands in the East are small—Granby referred to them as "islands of sanity," the same phrase Carlton Wagner had used quite independently. Old trees can be found in isolated pockets, usually on or near steep slopes that protect the trees from storms and make them too difficult to harvest, and that make the land unappealing for either agriculture or building sites. Such pockets, according to Bill, hold some very old trees.

> For example, there are hemlocks in the Berkshires that are approaching four hundred years of age. That's old growth! There are birches, yellow birches, that professional foresters [who have accompanied] me didn't even recognize [because when they are that old they look different]. Those trees [the birches] are between two hundred . . . and three hundred years old.

Bill defined an old-growth forest as one in which a number of trees have reached at least the halfway mark in their normal life span. For most species of trees common to New England, he thought a reasonable average would be around one hundred and fifty years. Of course these old trees would be outnumbered by much younger trees as well as early successional species. An old-growth forest is not, at least in the Northeast, by any means uniformly made up of old trees. That is part of what the advocates see as the beauty of an unmanaged old-growth forest. There are trees of many kinds and of all ages. What is just as important, in the rotting remains of fallen trees, new shoots proliferate. The floor of an old-growth stand is thick with decomposing organic matter that makes a congenial habitat for countless insects and small plants and animals. As Granby said: "It becomes literally a biological zoo there."

Ray Asselin had just come back from a trip exploring old-growth stands with Bill Granby when I interviewed him. He was struck by the differences between what he had just seen with Bill and what he remembered from his countless hours in the Quabbin. It is worth quoting him on the contrast at length.

> When you look at the Quabbin forest, or any other forest around here, it's young and it has small stems. They [the MDC] don't let them get too big. They leave some but they start harvesting when they are at a certain

size. And when they harvest, they take the entire tree . . . they take the saw log, they take all the limbs for chips. Everything is gone so the place, when they're done, looks like a park. . . . When you look at an old-growth forest it's a magic place. Things have had a chance over hundreds of years to grow and decompose and change and you get a very different look to the forest, very different structure. You have these massive old trees that have bent and gnarly limbs, and trees that are leaning over, that just have character to them. And you have fallen logs that may be huge and that will last for decades. . . . They become moss-covered and seedbeds for new trees. And they hold water; they're like sponges. They're soaking wet. You know, that becomes habitat for all kinds of other things, all sorts of things which are all part of a healthy forest. And a lot of that is missing from a place like Quabbin. They're manipulating the part of the forest that's aboveground and visible, as though that was all there was. But that's not the case. The underground portion is just as vast as the aboveground, and although we don't see it, it's just as essential to the health of a forest. I think by lopping off trees at the soil and removing all of that material and never allowing the forest to age is a horrible thing to do.

Bill told me that he knows of small enclaves of old growth whose location he keeps to himself. They are so fragile, so precarious, that he worries lest increased foot traffic from the curious destroy their integrity. These islands are, without stretching the point, tabernacles for him, places where time stands still and the presence of creation, past and present, weighs heavily.

I go to bed at night fearing that too much exposure [will] cause problems. . . . People will love them [old trees] to death. Old growth isn't something that excites large numbers of people but they [the old trees] won't take that much. Fifteen to twenty people going through some of these sites makes a difference. I have gone back [to a particular site] a year later and found my own footprints. . . . So what I try to do is I involve the best people I can . . . and let them be part of my conscience.

This sort of fragility is not what advocates of old growth celebrate, however. Quite the contrary. Central for the partisans is the stability and resilience of old growth. It is fragile only if humans intervene or disrupt things. Otherwise, left to itself, old growth is the organic equivalent of the mythical perpetual-motion machine. As old trees die they provide nourishment for young successors in what old-growth advocates variously describe as a process of achieving "balance," "equilibrium," or "climax." The whole is delicately fine-tuned to local conditions.

Ray Asselin added a jab at the MDC to his account:

An unmanaged old growth forest reacts as it has to to conditions and will grow whatever trees are appropriate for the conditions and the time. They [the MDC] are fighting that. They are trying to make the trees *they want* to grow grow in a place. And they don't have the wisdom—nobody does—to know that they're doing the right thing or the best thing.

In other words, nature knows best. Undisturbed forests are not only stable, they are models of efficiency. Echoing the nineteenth-century theme of nature's economy, advocates of old growth insist that there is nothing superfluous in an old-growth stand. Everything has a purpose and a place, even diseased trees. And, no matter how obscure the connections, everything makes some contribution to the viability of the whole. This is why human intervention is anathema: we don't know the connections and so we blunder in and throw everything out of kilter. Whether we attempt to suppress fires or control a blight or attempt to regulate a deer population, we are altering the natural processes and thereby disrupting an entire system that would otherwise balance competing organisms and produce an optimal mix.

Balance is a crucial concept for the MDC's critics. Old growth is the epitome of a self-adjusting, infinitely flexible system in dynamic balance. Each constituent element exists in harmony with the other elements in the system. In this sense, old growth or undisturbed nature is taken by many of the critics of the MDC as a model for how we humans ought to see ourselves in relation to one another and to nature. Bill Granby said:

[T]here's a certain peace that you can get in these old places. Because there's a sense of balance. . . . Everything is in some state of balance. The amount of material that goes back to earth comes back in regrowth. And I believe when people walk back into those places, it touches something primitive inside of them. Primitive good, not primitive bad.

If we could relinquish our need to dominate nature, if we could let nature "regroup," a new era would be ushered in, an era in which peace and harmony would come to prevail. Peter Gomes was at his most intense when he said that if we would only control ourselves, control especially the growth of our population, we would be sitting pretty.

> The forest, in all its lushness, propagates people. If you just remove it [the forest], you are removing your womb. . . . Somehow we have to get back and let nature have some room so we can continue to live. . . . Maybe this sounds ridiculous but I think human beings have a chance for utopia, to live in a Garden of Eden. If we just realize our situation, stop producing people, and . . . let nature recover.

Sitting on Peter's porch on a pleasant sunny summer morning, over-looking an undeveloped stretch of scenic river, it was easy to see how he could hold on to this conviction. But Peter is far from being a starry-eyed optimist. He and most of his fellow critics were in fact pessimistic about the future. Most shared Peter's dismay at the increase in world population (Rick Prevost was the only one among the critics I spoke with to insist that it is our attitudes and values, not our numbers, that is the problem). Indeed, eight of the nineteen critics I interviewed indicated that they had chosen to remain childless because of concern for the environment. Martin Dodge thought it was "suicidal" to back away from population control and yet, with touching ruefulness, wondered whether the decision of environmentally conscious persons not to have children would mean that "conservation might get bred out of the human race."

This was not the only apocalyptic vision worrying the critics. Bill Granby saw environmentalists like himself locked in a losing battle to preserve what little remains of our natural heritage. Lenore James and her husband, Steve, admitted to being gloomy and "pretty pessimistic about things in general." "There are so many people," Lenore observed, "so many that are deprived, that there's no way our planet can really fulfill all of the needs." Dorothy Reading reported that she was "not very happy living here, in this time. . . . I think we're all pretty much all committing suicide at this point." Ray Asselin is optimistic whenever he is learning more about nature and contemplating its intricacies, but "there's the other side where you see this sea of humanity overwhelming everything and consuming everything." Lorraine Near wills herself to feel upbeat, "I have to get up every morning and believe that things are getting better," but moments later she worried aloud that there may be "ten times as many people as this earth can sustain comfortably." Joan Williams confessed similar fears.

> I don't have much hope. I mean I really don't have much hope for . . . the world. I feel really scared . . . and when my children say 'I don't think I want to have any children' I say 'I understand.' . . . I think our life will live out the next thirty years or so and we'll probably be kind of okay. I think our grandchildren are not going to be okay, it's going to be really awful.

This pervasive sense of impending disaster lent urgency to the critics' efforts to change course at the Quabbin. The larger world may be spinning out of control but here, virtually next door for many of the critics, there was a chance to turn things around. At stake was not only the Quabbin but a sense of historical moment. If management could be scaled back at Quabbin, everyone would be able to see that the answer to our environmental problems isn't always more intervention. Here was a prime opportunity to break the addictive cycle of using ever more intense interventions to try to cure the ill effects of past interventions. The vision of a large area being allowed to heal itself and being restored to its former harmony, a monument to hopefulness, made the critics resolute, despite their pessimism.

3

Taking Care of Nature

> The history of civilization details the steps by which men
> have succeeded in building up an artificial world within the
> cosmos. Fragile reed as he may be, man, as Pascal says, is a
> thinking reed: there lies within him a fund of energy, oper-
> ating intelligently and so far akin to that which pervades the
> universe, that it is competent to influence and modify the
> cosmic process.
>
> T. H. HUXLEY, *Evolution and Ethics*

THE METROPOLITAN DISTRICT COMMISSION made an easy target
for the critics of the hunt. Since the Quabbin is completely contrived, a
point the critics were willing to strategically "forget" when it suited
their needs, everything on the reservation, good, bad, or indifferent,
could be laid at the doorstep of those who had done the contriving.
Though the MDC staff most often operate well out of public lime-
light—either in the woods of the Quabbin or in offices nestled into a
functional brick building near Winsor Dam—they are public servants
whose decisions have enormous consequences. For this reason, the
managers of the Quabbin, though they would rather be in the forest or
in the research libraries they use to keep abreast of shifts in thinking on
environmental management, are prepared to subject themselves to
public scrutiny. But they were clearly unprepared for the kind and
intensity of criticism they received.

This was not a matter of naiveté so much as a feeling of deep misunderstanding and betrayal. The MDC professionals generally shared most
if not all of the environmental ideals of their critics. At least as devoted to

the land and woods as their critics, the MDC staff also lamented what is happening to the larger environment. As we shall see in detail shortly, they shared the same broad sense that calamities surely await us in the not-too-distant future if we do not quickly cease abusing nature. Moreover, MDC staff freely admitted that the track record of environmental and resource management is at best mixed. They were acutely aware of the limits of their knowledge and of the potential risks of error. If they could have brought themselves to believe, with their critics, that nature could take care of itself, they would have been greatly unburdened. But they did not think that nature alone could take care of itself, much less yield, uncoaxed, the resources that a hungry and thirsty world needs.

For the most part, the same can be said of the MDC's supporters. With the exception of some of the hunters I interviewed, none of the proponents of the hunt were unabashed enthusiasts for attempts to improve upon nature. Nor were the nonhunting proponents of the hunt particularly well disposed to hunting or hunters. It is striking how few and narrow were the issues that divided those for and against the hunt. This is not to say that the division occurred over trivial matters. Indeed, at the core of the controversy was nothing less than a disagreement about the nature of nature.

As we have seen, the critics typically portrayed nature as tending toward balance: if we could just keep our hands off, equilibrium would be the outcome. Of course, keeping our hands off implies that humans are, in some important respect, outside nature, that we introduce unnatural forces and otherwise upset natural rhythms. The proponents of the hunt, by contrast, rejected such notions of balance and equilibrium. They were much more likely to characterize nature as chaotic, given to long oscillating cycles within which disturbance and disruption are commonplace. Moreover, they resisted the temptation to see humans as foreign elements in an otherwise harmonious scheme of things. Humans, from this perspective, are just one among many of the sources of disturbance and disruption. Our effects on the environment are no less natural than those of the deer. The difference is that we can become conscious of the effects we are having whereas the deer cannot. That is why we can manage the deer, not vice versa. Of course we *can* manage deer. The question remains: *should* we?

For the men who worked for the MDC and their supporters, the answer was obviously "yes." The deer herd was simply too large to be sustained without doing serious, perhaps irreparable damage to the watershed. For the proponents, water was the bottom line. In fact, water is the reason, the only reason, the Quabbin exists. It is not an experimental station, even though experiments and studies are being done there all the time. Thomas Conuel might have come closer to the truth had he dubbed the Quabbin an "accidental laboratory" rather than an "accidental wilderness." It certainly is more the former than the latter. Ironically, it is experiments like the introduction of eagles that have contributed to the popular idea that the Quabbin is a wilderness area.

The MDC and its supporters were unanimous, even adamant, on the water issue. While sympathetic to the principle that areas ought to be left untouched or, where possible, restored to more nearly their natural state, the MDC faction, as I shall refer to the men and women who supported the deer reduction program, could not understand the critics' willingness to let the Quabbin alone "to see what happens." As much as they enjoyed the wilderness aspects of the Quabbin, the MDC faction assumed that the controversy over the deer was not about wilderness but about keeping the water supply of the Boston metropolitan area safe and reliable. Robert O'Connor put the matter bluntly. Reflecting on a recent vacation in the Adirondacks, where he and his family had savored the joys of a minimally managed wilderness area, he observed of the Quabbin:

> When you look at what's at the bottom of all this land, which is this artificial reservoir, and we sit right here on top of this dam that holds it back, that is what we are concerned with: protecting the water. I mean that's even the name of our division, Watershed Management. We are mandated by law to "protect, preserve and enhance" the watershed. . . . If the valley was still here and the state owned the land, it would probably be very appropriate to have a wilderness area. The "lake" [makes it seem more] like a wilderness landscape but really, when you think about it, it's totally different from what a wilderness would be. To create two huge dams, the biggest ones on the East Coast almost . . . and create an artificial lake for people to drink out of. . . really has to change how you look at it.

Elisa Campbell, a lifelong environmentalist and a force in the local Sierra Club, made it clear, both in her public presentations on the panels the MDC held as well as in the interview she granted me, that water had to be the decisive consideration in decisions about the Quabbin.

> If this were a wilderness area, I would say leave it alone. Then whatever happens happens. It is *not* (her emphasis) a wilderness area, it is a drinking water reservoir. And I think the people who manage it have an obligation to try to maintain continuity. . . . They can't afford to run . . . risks.

We shall explore this notion of risk shortly, since it reaches to the marrow of one of the main bones of contention between the opponents and proponents of the hunt. For now, it is sufficient to note that while the critics saw the deer as either benign or merely nettlesome, the proponents saw the deer as appreciably adding to the stresses on the watershed. The MDC was powerless to control some of the stresses: acid rain and air pollution were mentioned by almost everyone as real and growing dangers. In the face of such perils as these, it was not prudent to ignore the stress deer were imposing on the forest. From the MDC staff's point of view, any mistake on their part would not only create havoc for the state's consumers of water but might also ruin their professional reputations and even cost them their jobs. Unlike either their critics or their supporters, the agency staff felt they had little room to maneuver. Critics could talk about waiting to see what would happen; the MDC staff, many of whom were curious themselves about what would happen if the deer and the forest were left to their own devices, did not feel they had the luxury of such free-wheeling experimentation. Too much was at stake.

As public servants charged with providing a safe and secure source of water, they had to set aside their personal feelings and follow the management policy most likely to reward the public's trust. This placed many of the advocates of the hunt, both MDC officials and their public supporters, in the awkward position of endorsing a move that would allow hunting, albeit tightly controlled and heavily monitored hunting, but not cross-country skiing or trail biking. The irony was not lost on either supporters or detractors of the hunt. Everyone I interviewed understood and respected the MDC's desire to restrict public access in order to protect the watershed from contamination or mischief. Those who walked and communed in the Quabbin did so reverentially, understanding that it was a great privilege, not a right. It irritated both critics and supporters of the MDC that hunters would now have access. Even some of the hunters expressed ambivalence on this score, so broadly shared was the feeling that the Quabbin was special and needed to be protected from undue human pressures.

The managers and their allies were, in other words, uncomfortable with the course they themselves were recommending. While some of their critics portrayed them as supporters of resource consumption and manipulation of the environment, most of the proponents were in fact less than enthusiastic about intensifying human intervention and increasing the human presence on the reservation. The MDC had, with utmost reluctance, acceded to pressures to open the reservoir to sport fishing in the 1950s. The prospect of another foot in the door was not a welcome one. And yet, there appeared to be no choice: the deer had to be controlled.

Reluctant Managers

The critics of the Metropolitan District Commission, as we have seen, were inclined to portray the MDC and its supporters as zealous managers, addicted to meddling with nature. This turns out to be a considerable oversimplification. Not only were many of the proponents of the hunt uneasy about the increased public access that would come with the control of the deer herd, the MDC staff and their environmentalist allies were themselves ambivalent about management. They were well aware of the errors of omission and commission that resource managers, themselves included, had made. While stressing the importance of science, they freely acknowledged that the science they depended upon to guide them was still young and that there was much that was not known. They were very conscious of the risks entailed in any management decision, risks that were made more daunting by the near certainty of unanticipated consequences of their interventions. How would the deer respond? Would something else move in to fill the vacuum created by a reduced deer population? Would this something else be innocuous or even more insidious than the deer? Would it provide more ammunition for the critics of environmental management and more bad publicity for a public agency understandably concerned for its autonomy in budgetary and personnel matters? Water lay at the heart of the matter, but the veins and arteries of the policy planning involved the livelihoods, self-respect, and professional identity of the people in charge of maintaining the watershed.

To get an idea of the complexity of the issues involved, consider the beaver. The beaver was essentially trapped out in Massachusetts by the time of the Revolutionary War. How beaver found their way back to the Quabbin remains unclear, but they appeared on the reservation in the early 1950s and enjoyed a rapid increase, mainly owing to the large amount of new growth on the watershed. The wildlife biologist for the MDC, Paul Lyons, whose graduate work was done on the beaver, related what then occurred.

> They [the beaver] showed the classic sigmoid growth curve where they start off fairly slowly and then have a very rapid growth phase. And then, oftentimes, they actually overshoot the carrying capacity of the habitat. They crashed substantially from a high point in the late seventies and early eighties to the present time and are now about a quarter of their peak population. . . . But then they rebound again usually, as the habitat conditions rebound after the population crashes. But because of the influence of the deer, the habitat is not really rebounding.

Deer and beaver ordinarily have a symbiotic relationship. The beaver in effect clear-cut an area and then, when they are out of food, move on.* In their wake, new growth begins, which is ideal browse for deer. If the deer are not too numerous, regeneration continues until such time as the area once again becomes ideal for beaver and in they come, starting the whole business over again. But in the Quabbin, the deer are so numerous that regeneration is not occurring. This holds the

---

*The beaver cleared about twenty-five hundred acres along the stream beds and upland wetlands of the reservation. The clearings have remained open because of deer browsing. By contrast, the cutting by the MDC that the critics have so roundly condemned totals approximately five hundred acres.

beaver in check, but it has also forced them to modify their eating habits. The result is not a happy one, though almost no one mentioned it either in the public meetings or in interviews. Only two people, one an MDC forester and the other Tom Berube, the president of the Massachusetts Council of Sportsmen, pointed to the beaver as another potential problem for the watershed. Berube was convinced that the beaver posed a greater hazard than the deer in the long run. Whether the beaver will affect water quality or not remains uncertain. But the beaver-deer interaction has already had a subtle deleterious impact on the Quabbin forest.

Bruce Spencer, the head of MDC forestry and the target of considerable hostility from the ranks of the critics, is a tall and engaging man. He is plainspoken and quick to make clear that he loves trees—and not just the commercially attractive, tall, straight trees his critics accuse him of fostering. As he described the forests of the Quabbin, he noted with exasperation that the beaver had clobbered the only really old growth in the reservation.

> The only prehistoric plant communities of any size that I know of on the Quabbin were the wetlands, remote wetlands, that they didn't clear for agriculture. . . . [W]e had these very old tupelo trees that were three to four hundred plus years old. . . incredibly dense thickets. And the beavers came in because they weren't being trapped and their regular areas were exploited . . . and they wiped them out. . . . Took 'em out. They flooded 'em, killed everything.

Reducing the deer herd will, in all likelihood, give the beaver more room and could mean that their numbers will rise to unacceptable levels. That would raise the prospect of trapping to control the beaver population. Trapping would be yet another "hot button," exposing the MDC to more unfavorable publicity and political pressure. In very real terms, the managers of the Quabbin felt trapped themselves between the endless challenges that nature posed to the water supply and the equally endless string of criticisms that inevitably resulted from their attempts to respond in turn, to both nature and political pressures.

In the public airings of the deer problem, the MDC had little to say in response to the barbs aimed at past management practices. Staff members were not in the least inclined to defend everything the agency had done in the past. Without being defensive, MDC spokesman Robert O'Connor kept trying to focus the discussion on the size of the deer herd and what to do about it, not on a rehash of who or what was to blame for the high concentration of deer. This is not to say that the MDC accepted responsibility for the number of deer. The agency claimed that the deer herd would have been excessively large even if it had done no logging whatever. The coincidence of low predation and the overall age and composition of the forest when it became a reserve was ideal for creating a booming deer population.

Adding to the MDC's exasperation with its critics was a sense of betrayal. Some of the most vocal detractors had earlier been equally vocal in supporting MDC management practices. The activity most frequently singled out as evidence of the MDC's ineptitude was the clear-cutting done in the 1980s. As noted earlier, the clear-cutting was done in order to increase water yield at a time when Boston officials were seriously contemplating a diversion of the Connecticut River to augment the Quabbin's water sources. When the water yield was increased and demand was reduced through conservation measures, the need for a diversion disappeared. At the time, the cutting of the red pines was lauded by Massachusetts Audubon. One MDC forester drew the connection between the diversion controversy and the current struggle over the deer.

> People pulled their hair out [over the proposed diversion]. Said "No, by God you'll never do that!" And those people who were pulling their hair out 'cause they didn't want that diversion came to us and said clear-cut the red pine . . . as an alternative. Increase the yield. The same people . . . today, when we're talking about deer and forest management, are coming at us saying "Why did you ever make the decision to clear-cut? You know it fed the deer, you know it's a terrible practice."

Perhaps the clear-cutting had been a mistake, but it was a mistake that was widely supported and one that was believed by one and all to be a far preferable alternative to what was waiting in the wings. Rather than defend the past record, the MDC faction insisted that the agency had consistently acted with good intentions and made conscientious efforts to study and understand the dynamics that linked forestry practices, deer, and water supply at the Quabbin.

There was some poignancy in the way the foresters spoke about their work. Thom Kyker-Snowman, who spent years as an instructor in Outward Bound before returning to graduate school to earn a degree in forestry and begin working as a forester for the MDC, talked about how he felt when trees are being cut:

> I mean I have a terrible time when I go out to check on a logging job. I still have a hard time hearing the trees crashing down to the ground. I'm not crazy about seeing a beautiful big tree go from this (expanding his arms upward) to something inert on the ground.

Bruce Spencer tried to explain his fascination for the process of plant succession. The sheer power of the drive to regenerate, the idea of seeds lying dormant for sometimes hundreds of years, just waiting for the opportunity to burst forth, was overpowering for him.

> I'm still in awe. I mean all those seeds . . . ready to happen, waiting for a disturbance. I mean that's such a powerful force. I'm always in awe of that.

That deer could be allowed to run rampant, their numbers out of control, nipping all this plant life in the bud was incomprehensible to him. While critics portrayed the MDC foresters as insatiable destroyers of woodlands, the foresters themselves talked lovingly about trees and said how upset they felt when they surveyed the damage the deer were doing.

The MDC's supporters also made plain their regard for the forest. Elisa Campbell was characteristically straightforward on the matter: "I found out that when it comes down to right to life, deer versus trees, I'm with the trees." This sort of regard for the forest had led some in the MDC faction to shift from enjoying the scenes of deer moving with grace and serenity through the reservation to seeing the deer as vermin—"horned rats," one forester dubbed them. Kyker-Snowman, only slightly more charitably, recalled that "Somebody wrote up a little thing from the perspective of a small tree, an acorn that had just germinated and had its first leaves out, and saw this huge ungulate bearing down on it and chomping it." He went on to characterize deer as "charismatic megafauna" whose beauty blinds us to their more insidious potentials. Bruce Spencer confessed that having observed the damage the unchecked deer population had done on the Quabbin had converted him from an opponent of hunting to a hunter. Though he had yet to kill a deer and was more than a bit self-conscious with his about-face, he nevertheless relished the thought of doing his small part to help keep deer populations within environmentally sound limits.

Such dismay at deer damage carries with it an acknowledgment that the deer are present in large numbers because of the way humans have shaped natural forces: by clearing the land, eliminating predators, and then forbidding hunting, we have created an ecological vacuum in which the deer are certain to multiply. The MDC faction believed that, having created this situation, we have no choice but to intervene again in order to save the forest from the ruinous activity of the deer. Though the MDC faction might have resented the pejorative connotations of the critics' charge that they were addicted to manipulating nature, they took it as given that our long history of transforming nature has left us no alternative but to manage. Kyker-Snowman again:

> We're so far into manipulating already, we've done so much to manipulate it [nature], that we'd better recognize that only through science and constant research can we learn what will take us in a bad direction and what will take us in a good direction. We have to bring nature into the directions that we need it to go. . . . You know there's the extreme point of view that says "let nature take care of the problem," but one of the things that goes along with that, and I'm not necessarily opposed to this, is that there would need to be some drastic reductions in our own population to correspond with that. Unmanaged nature can't sustain us.

Alexis McNair has been in love with the Quabbin for over twenty years. She came to the western part of the state years ago for a conference and, seeing a large unmarked chunk on her road map, decided to drive along the perimeter on her way back to Boston. She recalled the moment for me:

> And I stopped there, it was one of those cloudy afternoons, in front of the Visitor's Center. . . . It looked incredibly Scottish, it looked like a tarn, and I was just smitten to the soul! I said this is the most beautiful place there could be.

Before long, she and her husband, both of whom have hiked and camped and canoed all over the United States, relocated to western Massachusetts, where she became an active member on the citizen's boards involved with the Quabbin as well as a leader of local environmental groups. An unabashed lover of wilderness who is skeptical of all assertions that humans can improve upon nature, Alexis nonetheless observed with a tinge of resignation that "once you've started [managing] there really doesn't turn out to be much of an alternative" to continuing to manage.

Charlie Thompson, a forester who works for one of the professional forestry associations in Massachusetts and who was on the MDC public panels as a "representative forester," went further. Like a number of others in the MDC faction, he emphasized how much humans have altered nature, intentionally and unintentionally.

> [There isn't even] a square meter of the earth that is now wilderness in some fundamental way because everything is subject to these [human] effects, sometimes invisible ones like ozone depletion. That's terrifying to me.

Charlie concluded, in much the same tone of resignation as had Alexis, that "it's impossible to leave nature alone." Management, in other words, is far superior to heedless or unconscious activity, or to romantic notions of leaving nature to its own devices.

This idea was couched in broader terms by Elisa Campbell, who, along with Alexis, was one of the most vigorous local exponents of the view that environmental protection should be put before human convenience. Like virtually everyone in the MDC faction, she wanted to see as much wilderness preserved as possible, for its own sake as well as for the sake of species that require deep woods and minimal disturbance by humans. But that didn't mean an end to management by any means. She explained:

> We need a lot of habitat. And some of that we can probably best get by some careful management. It has to be done by people who recognize that they have a lot to learn and that they don't want to go overboard in one direction or another because we don't know that much. . . . We've already removed all the predators and now the hoofed leaf eaters are destroying the environment, the habitat, for everything else. Human

beings and their grazing animals have created deserts everywhere that they've been and we're continuing the process. We haven't learned a damned thing. So it's not that I have great confidence in human management. But I also think that at this point we're so dominant and we've destroyed so much habitat and paved it over . . . that I think we have a responsibility to try to provide some variety of habitat.

As these comments make clear, the MDC faction shared its opponents' dim view of human behavior—overpopulation, wasteful consumption, the unchecked use of hazardous chemicals. But they drew from this dismal scene a conclusion diametrically opposed to that drawn by their critics: we must manage more and manage better. It is as if we are condemned to manage by our own past and current follies, much as Adam and Eve were condemned to labor for their sins. The only way to even approximate the "garden" now is to be as thoughtful and prudent in our manipulations as we can be. "Thoughtful and prudent" is just how the MDC faction saw the efforts to manage the Quabbin. As we have already noted, the defenders of the agency accepted the fact that the Quabbin was a reservoir and that it was never intended to be anything but a tightly controlled and managed watershed. The supporters of the MDC spoke appreciatively of how much the agency had been able to do, within this mandate, to create and maintain a unique and beautiful area. The following remarks by Alexis McNair capture nicely the shadings of difference between the two camps:

> I agree with the 'no management' people that we don't know what we're doing, we haven't the slightest idea. So I think that our instinct, which is that there should be some undisturbed areas, is absolutely correct instinct. Whether it arises aesthetically or spiritually or from a kind of gut sense that most people have who study the natural world. . . . So I'm not crazy about timbering. If they never cut down another tree I'd be happy as a clam. But I do think that it [the Quabbin] is probably as well cut as any place, anywhere I've been. [I've been taken to] places which were cut . . . five years ago. You just cannot tell [that trees have been taken out of there].

Elisa Campbell was more adamant in her defense of the MDC's management practices.

One of the same people who is now violently objecting to the cutting [of trees] or the deer hunting, is a full supporter of the loon and eagle restoration program. In the name of nonmanagement! Give me a break! . . . They have this fantasy of . . . real wilderness. Well it's not a real wilderness. It has never been treated like a real wilderness. . . . It's not fair to blame the MDC for not planning for a wilderness area . . . that's not what they said they were doing.

Elisa had also been taken through areas that had recently been logged and found it hard to see any trace of harvesting. She had no doubt whatever that the MDC's logging operation was as sensitive to environmental concerns as it could be.

Twinkling eyes and a quick smile belied Charlie Thompson's angry words when he reflected on the critics' charge that the MDC was embarked upon a commercial logging strategy.

If there's anything that can make me lose my temper and lose my cool, it would be that discussion. . . . It's been asserted that the MDC is "in bed" with the industry. There isn't even an industry in this state! There are no big [logging] companies. It's all the equivalent of the family farm. The [loggers] have been at war with the MDC for twenty years because they won't allow the good trees to be cut. They just keep growing them. You know, "good" from the loggers' point of view.

From his perspective, the MDC logging bore no resemblance to commercial logging. And while Charlie, like the rest of the MDC faction, wasn't overjoyed with logging of any kind (moments earlier in our conversation he had said that "when I go for a walk in the woods when I'm not working, the kind of places I go are the places that are wild, that have big trees"), he knows that we require lumber and other wood products. While defending the MDC's quite uncommercial operation, he asked rhetorically,

what is wrong with growing straight trees? I mean in the eyes of God I don't think there's any difference between the straight tree and the crooked tree . . . so what's wrong with growing a tree that has commercial value? What's morally wrong with that? To me, it's morally right. I mean not that every tree has to have that quality. . . . But you know people consume a certain amount of wood in their lives and if that tree had come from twenty miles away rather than five hundred or two thousand, that is a great good.

If the Quabbin yields a certain amount of quality sawlogs, as part of an overall plan to maintain both the habitat and the watershed, so much the better. Every log from the Quabbin represents board feet that New Englanders don't have to expend diesel fuel to secure. It also takes pressure off private stands of forest where the standards are primitive by comparison to those the MDC requires. The cutting on private lands is far more intense. Bruce Spencer estimated that were the Quabbin to be managed as a commercial woodlot it would yield annually "at least three to four times what we're doing, on a conservative basis."

More is involved in the difference between what goes on at the Quabbin and what goes on in most forests. The MDC foresters carefully stipulate which trees may be cut. They also insist that the loggers use specialized equipment designed to reduce to an absolute minimum the amount of collateral damage associated with felling the trees and removing the logs. And they require that chain saws and other equipment be refueled on pads of the same absorbent material used by the Environmental Protection Aagency in cleaning up oil spills. It is small wonder that, as Thompson pointed out, the loggers have fought with the MDC. Such standards raise the costs of cutting timber far above what would be entailed in most logging operations.

The MDC foresters were understandably proud of the fact that their forestry practices were exemplary, so far above those of commercial loggers as to not warrant being spoken of in the same breath. Kyker-Snowman was particularly insistent on this.

> Because this is a watershed, because we don't have any commercial pressure, because the state has not said you must produce X million dollars from timber cutting every year, we have the luxury of demanding the absolute highest standards in timber harvesting. I would say the highest anywhere in the country.

From this point of view, the Quabbin could be regarded as a model of how prudent management can come close to meeting multiple and what are, in some contexts, even mutually exclusive goals. We may not ever have our cake and eat it too, but, the MDC faction insisted, we can have ample potable water and also have a forest that protects the water, affords modest quantities of commercial-grade timber, and sustains a remarkably rich and varied biotic community that can inspire and soothe, as well as recruit for enlightened environmentalism.

The MDC faction did not see the management of the Quabbin as inimical to environmentalism; it *was* environmentalism. In fact, members of the MDC faction, in the aggregate, had been far more involved in environmental activism, both with respect to the Quabbin and in much broader national and international contexts, and were more active in prominent environmental organizations than their opponents. They were also at least as likely to be enamored of wilderness, both as an abstract ideal and as a legitimate goal of public policy. They were clearly as committed to seeing more areas set aside for "wilderness" as were their critics. Indeed, those in the MDC faction had logged more days in remote wilderness areas than their counterparts on the other side of the deer controversy.

It was, then, with some justification that the MDC faction viewed many of its critics as "johnny-come-latelies"—well-intentioned, but inexperienced and easily swayed by simple-seeming solutions. Having been activists for a longer span of time, and having had to confront many complex issues that often involved gut-wrenching trade-offs, members of the MDC faction were impatient with what they saw as the naiveté of many of their critics. This added to their more general sense of exasperation. They felt that they and their critics should be allies in a common struggle for environmental sanity. They resented being characterized as part of the problem.

Referring to a recent instance in which some environmentalists were urging a return to unmanaged nature, in this instance by dismantling hydroelectric dams to restore rivers to their former wild state, Kyker-Snowman expressed this exasperation well:

> Environmental groups are jumping on the opportunity to challenge relicensing of hydro-dams to let the rivers run wild again. I hate the dams too but right now we have 'X' demand for power and that's not going to suddenly change. Hydro is "free" power once you've made the capital expenditure for the dam, and it's nonpolluting. So if you drop hydro, we'll rely more on coal and oil which are nonrenewable, expen-

sive and cause acid rain. And there's always nuclear in the background. So that's where some of the well-intentioned environmental efforts are taking us—to nukes. It drives me crazy.

The MDC faction regarded its opponents as oversimplifiers and romantics. The animal rights activists were the easiest to target because their single-minded devotion to the deer, to the exclusion of all other considerations, made them seem almost silly, certainly not serious about ecology or the environment as a whole. But as we shall see in greater detail in the next chapter, all but the hunters within the MDC faction expressed sympathy for at least the ethical concerns of the animal rights advocates. The MDC faction also considered the proponents of old-growth forest as naive. The crux of the matter, and the thing that most clearly divided the MDC faction and its opponents, was an underlying disagreement about the nature of nature. As we have already seen, the opponents of the hunt stressed the "balance of nature." If we let things alone, they believed, nature will work things out. By contrast, the MDC faction largely rejected the notion that nature will produce some preferable steady state of balance or equilibrium—nature, for them, was in constant flux.

Nature Is Disturbance

No one can pay even passing attention to nature without being deeply impressed by how intricately things are interrelated. Each element is connected, sometimes directly but more often indirectly, to every other element. For the critics of management and the hunt, these connections were defined in terms of balance. Things are fitted together, and when there is a disturbance the parts adjust, compensate, or adapt in ways that either restore the status quo ante or establish a new equilibrium. Harmony is the essence of nature in its pristine condition, according to critics like Gomes and Ventura. Thus, if we could only learn (or relearn, for those who believe that our distant ancestors once knew how) to live in harmony with nature, all would be well. To be clear, I am not suggesting that the proponents of such views are sloppy sentimentalists—everyone, on both sides and including the animal rights advocates, acknowledged that there was suffering, pain, violence, and death in nature. But for the critics, none of this was gratuitous: nature is economical. There is no waste, no unnecessary misery, no surplus that goes for naught.

By contrast, the MDC faction saw nature in constant change. Kyker-Snowman's view was echoed by nearly everyone in the MDC faction:

> My feeling is that if you look at the history of forests, especially in the Northeast, you will find disturbance is absolutely the norm. And disturbance of a pretty significant nature, including hurricanes.

Paul Lyons spoke to the same effect but put it a bit differently. When asked whether he thought nature could be characterized in terms of balance, he responded:

> What I think of as a natural balance most people would not think of as balance. I prefer to speak about equilibrium rather than balance because I think the word . . . allows for fluctuations which the word balance does not. Balance [suggests] almost a steady state. Equilibrium can entail a lot of fluctuations. If you look at the New England forest over the last four, five hundred years, for example, it's gone through dramatic fluctuations. But the whole time nature's kind of adjusting things, responding to disturbance. And disturbances happen with enough regularity that they preclude any type of true climax condition [the prototype of balance that the critics invoke] from really becoming established and maintaining itself. It may just about reach that point but then it gets hammered and knocked right back to an early successional stage again.

Nature, from this point of view, is unforgiving. Disturbances come in all shapes and sizes and they can have stunning consequences—among them sharp fluctuations in the populations of species, up to and including extinctions. To live in harmony with nature conceived in this way would be to accept wide swings of boom and bust, of feast and famine, that would be utterly incompatible with social arrangements dating back to early Mesopotamia. Nature is too chaotic, too unpredictable, to be trusted when the lives of millions hang in the balance.

Bruce Spencer pressed this sense of fluctuation and unpredictability further. Nature is indeed intricately woven of interdependent parts. Push here and you will get a response, but it may appear where you least expect or want it. Though interconnected and constantly adjusting to disturbances large and small, nature is not "managing." "What's nature's plan?" Bruce asked. He answered his own question:

> There is no plan. I mean, nature is just a series of events. The people [the critics of MDC's management of the Quabbin] think that nature's plan is benign, it's only going to turn out good things. . . . That really bothers me. As long as greed doesn't enter the picture I believe that what we are doing will protect the things that everyone's concerned about [old trees, habitat, biodiversity].

The MDC staff felt the burden of trying to anticipate distruptions most acutely. They had been entrusted with delivering water to several million consumers. The whole point of the reservoir was to store vast quantities of water against the times when rainfall and groundwater would be inadequate. Nature provides all right, but it doesn't necessary provide on time or in the places where it is most needed by humans. Harmony, balance, equilibrium, to the extent that these terms are useful at all, are useful only as abstractions. At best they are like averages. But as we all know, averages are deeply deceptive: though useful for some purposes, they necessarily obscure a very uneven and highly variable day-to-day experience. Even in areas of high annual average rainfall, there are droughts. Nature, abstractly conceived, can absorb these ups and downs. If some things perish in the drought, they will rebound once the rains come again or other things will take their place. But to call this "balance" is to miss the point, said the defenders of the MDC management strategy. Just because everything is interconnected does not mean that everything is a permanent fixture or that outcomes of disturbance are outcomes that we can live with comfortably. Elisa Campbell:

> Well, emotionally I would like to agree with the self-correcting idea but intellectually I cannot. Clearly, natural processes are bigger than we are and it seems to me we meddle with them at our peril. That doesn't mean that everything would be peace and harmony if we weren't here . . . or didn't have the technology to do as much damage as we're doing now. The processes are bigger than we are and they will continue but we may not like the results. I think it's romantic to think that nature is beneficial

to us, provides for us. . . . I mean people still think that we're the center
of the universe. . . . A lot of people have this image that nature is good
and we're mucking with it and we're bad. I think we're bad too, but that
doesn't make nature good, beneficial, or benevolent or anything.

In effect, the MDC faction, without pleasure or satisfaction, accepts
as fact that we cannot have a society even remotely like our own without substantial intervention and manipulation of nature. The conclusion they drew was not that we should back off or let nature take care of
itself. Rather, we need to improve our science, refine our management
techniques, and strive to minimize our burden on the regenerative
capacities of the earth. Charlie Thompson, who earlier in our interview
smiled wryly when he spoke of getting angry, furrowed his brow and
grew somber as he said:

> I smile when people say we should adapt to nature's way. I want them to
> tell me what that means in Amherst today. What do we do? It means
> stop driving your car immediately—that's petroleum and it comes from
> subjugating nature, exploitation. Do we shut off the lights? The ques-
> tion in my mind is how can we . . . be less intrusive, less impactful.

This is the venerable argument for stewardship. It would not be
stretching things too much to read the Quabbin dispute as a replay of
the argument between Pinchot and Muir at the turn of the century.
Then, as now, people wrestled with their relationship to nature—were
we part of or separate from nature? If we are part of nature, is everything we do "natural"? If we are separate from nature, should we regulate ourselves in order to conserve and preserve features of nature that
are important? Who shall judge what is important? Are snail darters
important? spotted owls? elephants? deer? Should we reduce our numbers so that other species can sustain or increase theirs? Should we
impose deprivations on ourselves so that other species can thrive?

Individuals within the MDC faction differed on some or all of these
matters in the abstract, as statements of personal philosophy. Some
were with Muir in their heart of hearts, and were drawn toward sympathy with their opponents who were arguing robustly that Muir's position become Quabbin policy. But that is where the Muir sympathizers
within the MDC faction drew the line: they could not accept the exten-

sion of Muir's preservationist philosophy to the Quabbin. The Quabbin was nothing but for the fact that it was a reservoir. It was and always would be a contrivance, an artifact of human design.

Paul Lyons, who, as much as anyone, wished to see our natural legacy kept intact wherever possible, reflected on this distinction:

> We've got to recognize that what we're doing is that we're making decisions based on our own personal value systems. Let's be clear about that rather than say we've got to reestablish a natural balance . . . because we've knocked everything out of kilter. I think that's hogwash. . . . We decide what's proper or not. Quabbin's a human-made resource basically. Even though it's a very natural system as well. It was made for human reasons and we may decide that in addition to water we also want to have forest products and we also want to have recreation and we also want to have bald eagles and everything else. Well that's all fine. Let's just agree on what those goals are going to be, those human goals, and then we can start talking about managing to bring them about. I just don't like to see this couched in terms of recreating natural balances because I just don't believe that's what we're really doing.

The issue, from this perspective, is not whether areas should be left in their pristine state, but rather what areas are appropriate for "benevolent neglect." A watershed on which several million people depend is inappropriate for the hands-off approach. Management, so long as it is conscientious and stewardlike, is absolutely necessary and unavoidable over large stretches of the environment. The MDC faction did not accept the allegation that the MDC had been irresponsible or unstewardly in its management of the watershed. On the contrary, as we have seen in terms of logging, the MDC faction thought the agency had behaved in model ways.

Having done their level best to protect the watershed from inappropriate human activity, the MDC now found itself with a deer population that was running roughshod over the reservation, damaging the forest at least as seriously as any number of prohibited human activities would have done. So why put up with it? Alexis McNair, though she found the idea of killing animals repellent, found the damage from deer even more intolerable.

If I saw this lack of regeneration on a piece of western land that was
being grazed by sheep or cows or goats I would say get those creatures
out of there. . . . Why should I feel differently because it's deer?

Implicit in such comments as these is a very different sense of balance or equilibrium. Good stewardship is about balance as a human
goal, not as a naturally occurring state. Like other dominant species, we
have the capacity to bend nature, making it accommodate to our activities. So do deer, and that is precisely what they were doing on the
Quabbin: the deer were also managing the forest. But they do not have
intentionality; nor do they have the capacity to fathom the consequences of their management practices. We have both, for better or
worse, and it is up to us to endeavor to maintain something like an even
keel, something that nature herself cannot do. Balance, in this view, is
not "natural," as the critics of the MDC argued; like so much else, it is a
human contrivance.

If balance, like beauty, is in the eye of the beholder, that gives us
humans considerable latitude, if not free reign, in determining what
species are in and out of balance, what the proper mix of species should
be, and what mechanisms are appropriate to keep things more or less
stable. As we have already seen, the critics of management rejected the
proposition that these matters are for us to decide. They insisted that
nature knows best what a forest should be and what animals should be
where and in what numbers. The MDC faction was much more comfortable making judgments about what a forest should be.

What Is a "Good" Forest?

Nearly everyone I interviewed agreed that the situation on the reservation was not "healthy." For the critics, the MDC was the source of the
illness. For the MDC faction, the deer were making the forest "sick." The
specific form this "sickness" took was *simplification*. The food preferences of the deer were determining which herbaceous species would
grow and which ones would be eaten to oblivion. William Healy, a
wildlife biologist with the U. S. Forest Service who has carried out
extensive studies of deer and other wildlife in the Quabbin, said the deer

"were simplifying the forest," just as the chestnut blight, Dutch Elm disease, and other pests have done.

> You've got one big herbivore that dominates things. The usual pathways of plant succession just aren't happening out there. The community has taken on a very different and unique character. . . . It reminds me of a European deer park, where predators have been excluded and large numbers of herbivores have been chomping for centuries. And that's fine for some, if you like the scenery and want to get rid of the understory and keep it open and parklike, but it's not my impression of what a forest ought to be. And if you're trying to preserve the flora and fauna, the native things that normally live there, it's not particularly good for that.

Simplification, the reduction of the range and variety of species, has been going on for a long time at the hands of humans too. By imposing our notions of what is useful and what is noxious, by privileging some trees, either for their utility or their beauty, and regarding others as "trash," some animals as "bad" and others as "good," we have simplified the ecosystem. Thus, predators like the wolf, the mountain lion, and the bald eagle were driven out in preference to livestock and "good" animals like deer.

Initially, the Metropolitan District Commission practiced its own simplification of the Swift River Valley ecosystem, first by clear-cutting the basin and then by planting large tracts of red pine on the periphery in monotonous orderly groves. They also initially showed a prejudice for deer, happy to have their help keeping the forest manicured and to feature them as an emblem of a highly stylized popular notion of the wild. The graceful, pacific animals, "charismatic mega-herbivores," became a source of favorable publicity for the Quabbin.

While the MDC discouraged public access to most of the reservation, it did set aside areas for visitors and, until quite recently, even allowed them to feed the deer. Dorothy Reading, an opponent of the hunt, recounted an experience that underscored the artificiality involved in promoting this sort of "nature."

> Just last fall I went for a hike near the observation tower. I was coming up toward the tower from below and I cut across an open area and there

were some people in the field and they were like shooing me away, like don't come through here. I thought "Who are those people?" This is a public area, a public trail. And I kept walking. And then I saw some deer. These people were furious with me because I chased away the deer. Then I learned that they had been putting corn out for days to attract the deer so that they could watch them. They said this is the animals' only sanctuary and I'm saying to myself, "Oh God."

People come to the Quabbin expecting to see "wilderness." As Clifton Read, the MDC's interpretive naturalist and the staff person who has the most contact with the general public, noted, most visitors are not aware of how thoroughly managed the Quabbin is.

There is this image in people's minds that Quabbin translates into wilderness . . . because there is a variety of wildlife and protected areas without buildings and . . . lots of vehicles. And in Massachusetts, with the possible exception of Mount Greylock, it is the closest we come to wilderness.

For some, like Dorothy Reading, wilderness means solitude. Others, though, think of it in ways that Read called a "zoo mentality." There is an orchard near the visitor's center that survived the clearing and construction of the dam. One of the former residents got permission to maintain the orchard, and it has become a magnet for deer. Clifton Read explained what happened.

Well, with all the apples, the deer came in. But then people started bringing Cheese Whizzes and pretzels and all this ridiculous food there and that's been one of the "activities." They go up there and dump it down on the ground and then sit back and they have their little lawn chairs and they watch the deer. We had to put an end to that.

No one I spoke with had any sympathy for Cheese Whiz nature watching, but many of the MDC's critics and supporters who were serious about nature study appreciated the chance to see wildlife, even while acknowledging the artificiality that might be involved. They enjoyed the opportunity to watch deer that were unafraid of humans. Alexis McNair drew a philosophical lesson from the animals' tameness.

It's been nice to be able to walk through there and see deer and stuff. You know, you get the feeling that the rest of creation regards us quite rightly as if we had AIDS. . . . They take one look at us and wwwhhhooooosh! And so it's nice to have a place where we are not revolting and dangerous to the animals.

Rick Prevost, the avid amateur wildlife photographer who wanted to let nature be so that we could enjoy "it as it was meant to be," was grateful for the number of deer, because it increased his chances of capturing them on film. Moreover, since the deer kept the understory down, nature buffs had a much more expansive visual field than they had off the reservation. In such a setting, not only were deer readily visible, other species of wildlife were also easier to spot and photograph. That deer are not naturally active during the day (in the reservation, because of population pressure, the deer have to forage for food throughout the day) does not seem to diminish the satisfaction observers report from seeing "wild animals in their natural condition."

The early forestry practices of the MDC were not intended to produce anything remotely resembling a "natural condition." Instead, they reflected a desire to stabilize the soils as quickly as possible to prevent undue runoff and siltation buildup. The red pine was ideally suited for this purpose because it is a fast-growing tree. But when the pines matured and the need for ground cover had been met, the red pines became something of an embarrassment. Critics of the MDC more than once referred to the red pines to buttress their contention that the agency's forestry was ill-informed and not to be trusted. The MDC foresters, of course, did not regard the red pines as a "mistake" or "ill considered." The trees had met an urgent need when they were first planted. But needs change and priorities shift, in watershed management as in life itself.

The MDC began to shift its forestry practices in the 1960s to reflect new needs, fire suppression among them. Concern for reducing the risk of fires led the agency away from pines and other softwoods. The intent was not to exclude these species from the watershed so much as to keep them from becoming predominant. Oak was likely to be the dominant tree species, ironically because of the past history of burning in the area. Oak thrives with periodic burning. Fire had long been used by Native Americans to clear land. The European colonists, up to the time the Quabbin was created, also used fire. Bruce Spencer described a common practice.

Some of their cutting practices were ruthless. When the sawmill men would cut the pine lots off and stack the lumber to dry, their immediate concern, since they had no insurance on the lumber, was to get rid of all the fuels. So they would burn the slash and often end up burning the whole lot. . . . That fire was the best thing you could have done for the oaks. Because that burned out all their competition. They have big root systems and they were just ready to go.

Given the need to suppress fires, Spencer is convinced that there will never again be the number of oaks on the reservation that are there currently. Though he was not addressing his critics' charge that he was managing the forest for commercially desirable oak, he made it plain that even if the MDC wanted to manage for oak, it could not achieve anything like the oak forest that presently exists.

Robert O'Connor reflected at length on the research he and the MDC forestry staff had done on the history of the forests of the region, research aimed at determining what the goals of forestry on the reservation should be. Initially, they explored notions of trying to create something akin to an "original forest," but they had no way to decide objectively when such a forest existed. Where do you draw your base line? Before the last Ice Age? After the Ice Age? How long after? Humans began exerting a strong influence on the forests of New England some ten thousand years ago. Where in the course of that history might you reasonably set a base line? Of course, some possibilities were easy to dismiss. O'Connor quickly ruled out the pre-Ice Age. Even if you could reliably know what the forests were like before the glaciers came southward, there was no hope of recreating them. O'Connor explained:

> As the ice was retreating and the climate was changing, the soil was developing [new qualities]. Initially it was kind of the northern tundra and then the forest evolved into coniferous forest. . . . Eventually, some think because of heavy burning, the oak type forest became widespread.

In short, the manifold changes that have occurred over the past several millennia have so altered the conditions that nothing approaching an "original forest" could be achieved even if the benchmark were to be drawn as recently as five or six hundred years ago. Our climate is different, the soils are different, the pests and pollution are different. The accumulation of large and small changes, some induced by human activity, some the result of uncontrolled natural processes, has altered the land and what it can and cannot support. O'Connor cited a recent Williams College study that revealed significant soil differences in otherwise similar areas that appear to be the consequence of whether the land was tilled for crops or left to pasture.

Some at the MDC had hoped that the Quabbin could be home to a "representative forest," a forest that contained the range of flora commonly found in central New England over the last century or two. Even this would be problematic, because of the impacts of several centuries of agriculture practiced in the Swift River Valley, not to mention the burning of which we have already spoken, as well as new blights (William Healy marked as very significant the blight that has decimated the chestnut, once a prevalent tree throughout the region and now virtually extinct in mature stands), and the stresses of acid rain, ozone depletion, and smog. O'Connor, with a mixture of scientific curiosity and dismay, summed up the whole enterprise in three words: "It's so complicated."

> If the habitat is not what it was, should we be trying to [grow] taller trees than might have grown here before? What should we do? Everything has been so disturbed and continues to be disturbed that leaving it alone now might be a very naive approach that people should think about a little more [than they seem to have].

The conclusion of this sort of speculation seems, in retrospect, almost foregone. O'Connor again:

> [It came down to] three basic approaches: unmanaged, even-aged management, or uneven-aged management. . . . It seems like uneven-aged management is the best approach, given what we are. If we were an Adirondack Park (with a very large land mass), then there's probably a real benefit to setting it aside. If you put all the pieces together, we're trying to protect what's left [to maintain diversity] but it's a very artificial situation. . . .

The virtues of uneven-aged management are the virtues inherent in spreading out risk over as large a population as possible. The MDC was publicly most concerned about the consequences of a major hurricane, a so-called hundred-year storm, like the hurricane of 1938, which blew down a large section of the watershed's forest cover. With the deer preventing new growth, the forest of the Quabbin has been becoming steadily more homogeneous, with respect to both the age and the variety of trees growing there. This simplification of the forest, added to the simplifications that past human activities have wrought, places the forest in greater jeopardy. A uniform stand of trees is more likely to be blown down completely than is a stand that has all ages and many types of trees.

Other factors as well were putting the Quabbin forest at risk. Bruce Spencer worried about the long-term consequences of acid rain and air pollution. If the amount and variety of new shoots is kept low by the deer, the genetic pool will necessarily be constricted and the forest will become that much less adaptive.

> We need younger trees. They are able to adapt and grow in [an altered] environment. You have tremendous genetic variability [in young trees] so that with lots of them you'll get trees that can put up with this change. . . . We're going to lose it all unless we address [the lack of regeneration caused by deer].

In diversity, there is resilience and strength. There is also stability, no small concern to people charged with keeping clean water flowing to millions of thirsty people.

But more than prudence and utilitarian considerations motivated the MDC faction. Aesthetic values were quite apparent in everything they said about the forest. Bruce Spencer:

> Jeepers, we can manage this place *and* have a beautiful forest, a nice place to walk. And other creatures can live here, even though some will be killed at times, but you know we can have something that's managed and be good. It can be enjoyed. Maybe they [the critics] don't want to find that out.

Elisa Campbell was incensed by the damage the deer were doing, not just to the forest's regenerative capacity, but to individual trees as well. Recounting a trip she and others had made with Bruce Spencer onto the reservation, she remembered seeing some small trees and thinking that since they were small, they must be young and that regeneration must be occurring. Bruce Spencer claimed that the trees they were looking at weren't young at all, they were stunted by the continual nibbling of deer. Elisa continued the account.

> And we said, "Aw Bruce, you don't know every tree in the Quabbin." So we nagged him into cutting one down. And it had at least thirty-eight rings in it [even though] it measured less than two inches in diameter.

She was clearly moved, imagining that tree and what it might have looked like had there been fewer deer.

The critics, as we have seen, supported the idea of diversity but claimed that the only diversity worth talking about was the "natural" diversity of an unmanaged old-growth forest. The MDC faction insisted that sound management could produce the same sort of diversity and with much less risk of the calamitous blowdowns and subsequent burn-off that they understood to be the "normal," that is, the unmanaged, fate of New England forests. Where their critics saw stability in old growth, the MDC faction saw constant flux.

Unlike the western old-growth forests, where trees can reach an age well beyond 500 years, forests in New England rarely remain upright past a 100 or a 150 years. Individual trees may survive longer than this, but New England's forests are continually being hit by tornadoes, hurricanes, and heavy, wet snowfalls that knock down trees right and left. These sorts of events would be of little consequence if no reservoir were at stake or if the major use of the forest were for firewood. With the need for wood and wood products constantly expanding, it made little sense to let trees in the Quabbin be ruined for most uses by the elements when they could be harvested on a modest scale. Moreover, and of vastly more importance for the managers of the reservation, letting nature be the forester put the reservoir in unacceptable peril.

The MDC and many of its supporters were confident that the foresters could produce every bit as much biological diversity on the reservation as could be had by adopting a hands-off approach. William Healy, the Forest Service wildlife biologist, personally felt that the forest of the Quabbin ought to be allowed to become a "typical New England forest." He went on to note that the deer, not the MDC foresters, were preventing this from happening. The deer were making the Quabbin "botanically boring" and, not coincidentally, much less good for bird watching. He noted that the forest stands on the reservation were nearing their peak age, as close to "old growth" as they historically get. He explained:

> In the Quabbin we know that over the last four hundred years, about every eighty to a hundred years a big hurricane has come through and blown down a lot of stands. So that on some sites out there you may get old trees periodically but the forest itself never [has gotten] much older than it is now.

Moreover, he challenged the contention that old growth was objectively different from the forest the MDC was aiming for. That there might be a subjective difference he did not deny, but that clearly was quite another matter.

> In my opinion, not much changes structurally in the eastern hardwood forest between the time it hits about a hundred and the time it becomes two or three or even five hundred years old. . . . There's nothing special in [an old-growth stand] for wildlife that you wouldn't have, say, in a forest that you cut down at age one hundred.

The only difference Healy could see between the forest the MDC was endeavoring to produce and an unmanaged forest might be in the "litter layer," the rotting leaves and limbs and trunks of dead trees. But he observed that the MDC was taking that into account.

> The amount that they [the MDC] remove is pretty minimal. One of the things that people don't realize is that you can move a forest along toward that old-growth structure. . . . The thinning that's going on out there, in my opinion, is actually accelerating the advance toward the structure of old growth.

All that stood in the way of achieving this was the deer. The deer, once thought to be an asset, had become a problem when the managers of the Quabbin shifted their managerial direction away from simplification and toward the promotion of diversity. Unfortunately, the deer were embarked upon their own campaign of simplification. After the MDC managers, the deer were the dominant species on the reservation, and they had been systematically frustrating the reproductive efforts of a whole range of herbaceous species and massively altering the forest. Alexis McNair, no great enthusiast for manipulating the environment, spoke for all within the MDC faction when she deplored the consequences of the deer browsing. "It's not a satisfactory forest," she said emphatically, because the deer were preventing a wide range of tree species from growing. "And when push comes to shove, I want to see a healthy forest growing back." The deer were shoving.

As these comments make clear, the MDC faction was quite willing to declare what the Quabbin's forest should be like. However much they might have liked an unmanaged old-growth forest in the abstract, they preferred to see to it that the forest of the Quabbin was as variegated as possible. The point was not just to pursue variety for its own sake, however. Variety of species and of age classes was a form of insurance against unwelcome and inevitable disruption. The more varied the forest, the more resilient it would be in the face of a major storm, an outbreak of disease, or an insect infestation. At bottom, the MDC faction was unwilling to declare that "nature knows best." Nature has to be coaxed, cajoled, and assisted in the maintenance of diversity.

As we have already noted, most within the MDC faction would have preferred to leave nature alone. But they knew that this was impossible. Few shared the unequivocal acceptance of management that most of the hunters expressed. As far as the hunters were concerned, the game and habitat management of the past fifty years had kept hunting viable, if not exactly ideal. Only a few of the hunters, though, would have agreed with the disparaging view of nature voiced by Tom Berube, a traveling salesman who devotes much of his spare time to sportsmens' interests as president of the Massachusetts Council of Sportsmen. Here is his sense of what happens when trees are left to themselves:

> I took a picture of a tree that was about a hundred years old, maybe 125 years old. It's the most pitiful, sorriest-looking thing there was. . . . Branches all over the place. This big around [holding his arms in a circle]. All broken, the tree all scarred to hell. Oh God. It's all ripped and there's blood coming down, the sap is running out of it. That ain't right.

Whether with enthusiasm for perfecting nature or with ambivalence and regret, those within the MDC faction accepted the idea that some management was unavoidable.

Where the critics portrayed nature as benign or even benevolent, the MDC faction saw nature as indifferent to outcomes: in Bruce Spencer's words, "nature has no plan." Supporters of the MDC plan, even those who did not like hunting and wished for a management tool that was less at odds with their personal values, shared the sense that humans have to avail themselves of nature's resources in order to live and that this requires stewardship. Stewardship means trying to preserve as much biodiversity as is consistent with intelligent resource use.

Put another way, we might say that our dependence upon nature, over millennia, has led us to manage and manipulate nature and natural forces. We have long since "lost our virginity," and nature's too. Now, our only realistic course is to bring as much prudence and knowledge to bear on our interventions as we can. We must create as much resilience as possible and distribute the risks as widely as possible so as to avert devastation. We have, from this point of view, an obligation to protect nature not only from wanton exploitation by humans but also from the wide and abrupt swings to which nature itself is prone. It is up to us to maintain "balance," and to determine what a "healthy" forest is. Even though there are almost certainly going to be conflicts about these matters, the conflicts should be settled on the basis of scientific understanding, not on the basis of fanciful ideas about nature or wishful thinking. In the final analysis, forests will be what we make them, not what nature in the abstract decrees.

4

Sport, Management, or Murder

AMBIGUITY AND AMBIVALENCE
IN MODERN HUNTING

What a coarse and imperfect use Indians and hunters
make of nature!
HENRY DAVID THOREAU, *The Maine Woods*

Looking back on my hunting days, it seems obvious that the
excitement and challenge of hunting was closely related to
the acquisition of food. There was never any question about
the morality of hunting, but neither was there any acceptance of killing for the sake of a trophy.
JIMMY CARTER, *An Outdoor Journal*

The aloneness of the hunter, and his thoughts of his hunting
past, are the very genesis of primitive energy. He is always a
young man, then, and making his most daring journeys. He
will not think of middle age, and even the responsibility of
his family will dim as he pauses, every sense alert for the
sound of what he plans to kill. This is really the only time he
is fully alive. All the rest is the dreaming time.
FRANKLIN RUSSELL, *The Hunting Animal*

THE METROPOLITAN DISTRICT COMMISSION would not have become embroiled in controversy, and its forest management practices
would not have been the object of intense and scathing criticism, had

95

the agency not proposed a public deer hunt on the reservation. Though few of the critics were full-fledged animal rights advocates, most were against hunting. Those who were not philosophically opposed to hunting held hunters in such low regard that the prospect of hordes of "low-lifes" loose with firearms on the reservation turned them against the proposed hunt. Except for the hunters among them, almost the same could be said for the MDC faction. I encountered no animal rights advocates on the MDC side of the fence, but there was little fondness for either hunting or hunters. The near-unanimous distaste for hunting and hunters among people who were otherwise at loggerheads about environmental management has complex roots in class and culture and the ways these interact to shape attitudes toward nature.

Hunting, at least in New England, has always been tinged with disreputability. Though modern hunters like to imagine the colonial period as a golden age, a time when game was extraordinarily abundant and hunting was celebrated as a manly art, the fact is that the early settlers, though dependent upon game for survival, fretted about the carnal pleasures afforded by the hunt. However practical hunting might have been, it was also fun. Not only was it fun, it was a threat to the precarious community the early settlers had established. Hunters could toy with thoughts of self-sufficiency; and if they thought themselves capable of going it alone, they could be tempted to resist the moral strictures and behavioral controls ordered by religious leaders and to leave the community, striking out on their own.

The conflict between Thomas Morton and William Bradford, only a few years after the settlers arrived at Plymouth, reveals the tension inherent in hunting.[1] Morton, obviously less devout and far less impressed by the "howling wilderness" of Bradford's imagination, left the confines of Plymouth to establish an encampment in the forest. There, he and several followers proceeded to hunt, fish, and trap, and to befriend and trade with the natives. They did more, in fact, than trade. They—God forbid—revelled together. To make matters worse, instead of being repaid for their debauchery by the Lord's wrath, Morton and his associates prospered. Their trade was brisk and profitable and their labors, compared to those of their more pious and timid counterparts, were light. It is not clear which was more threatening to the followers of Bradford, Morton's revelries or his success. In either case, Morton had to be stopped. Led by Miles Standish, a posse set upon Morton and his three associates, arrested them, and sent them back to England on the next ship. If the colonists had to hunt, Bradford and Standish wanted to make sure it would not be fun.

Hunt they did, and with a vengeance that betokens either enormous pleasure or great devotion to thoroughness. Economic historian James Tober, in his book *Who Owns the Wildlife?*, writes that by the end of the seventeenth century, less than a hundred years after the first settlement, the turkey had been rendered essentially extinct in Massachusetts. Deer, too, had become so scarce that authorities forbade hunting them for over two decades in hopes of restoring the herd.[2] William Cronon, the noted environmental historian, quotes Timothy Dwight of Massachusetts ruefully observing in 1790: "Hunting with us exists chiefly in the tales of other times."[3]

Throughout the colonial period and well into the nineteenth century, hunters pursued fur and feather, as much for the market as for their own larders. By the middle of the nineteenth century, furred and feathered creatures in the Northeast were under a terrible siege. Brightly colored songbirds were shot to provide feathers for the millinery trade. More substantial birds were shot for diners in the eastern seaboard's rapidly growing cities. Deer, rabbit, bear, and moose all furnished rural Yankees with a source of cash. To complete the carnage, the predators that competed with humans for these marketable animals (and which also preyed on domestic animals) were pursued relentlessly.*

Had this frenzy continued into the present century, there might well have been no deer left to create a problem at the Quabbin. But the slaughter started to stir the nation's conscience by the last third of the nineteenth century. As part of the more general move toward conservation and managerial stewardship that we sketched in chapter 1, socially elite hunters and fishermen, appalled by the drastic declines in fish and

---

*I do not mean to credit hunting alone with the depletion of wildlife that quickly followed European settlement. It is almost certainly the case that the habitat loss brought about by settlers' plows and axes did far more to reduce numbers and varieties of species than did their muskets and traps. Nevertheless, hunting and trapping for the market, an endeavor in which Native Americans were also enlisted, clearly took a heavy toll.

game, began to agitate for closed seasons, bag limits, and, most crucially, a strict prohibition on commerce in wild animals. Unable to sell
grouse or deer to the urban market, hunters reverted to the more
ancient incentives: subsistence and pleasure. Pleasure could be pursued
unabashedly by the late nineteenth century. Ironically, the pursuit of
pleasure afield came to be linked directly to restraint—pleasure was
enhanced not by the number of animals killed but by the self-imposed
rules of fair play that produced an aesthetic and stylized encounter with
wild animals.

Shaped by upper-class sensibilities and resonating to an increasingly
educated and urban population, the ideal of the sportsman began to
take hold. The ideal was promoted by a rapidly expanding group of
monthly magazines devoted to a readership eager to know about new
equipment and techniques, and hot spots for fish or game, and to read
tales of others more lucky than they might ever hope to be, the people
who could travel to Africa or Alaska in search of trophies and adven-
ture. The essential elements of the sportsman ideal included etiquette
and respect for game laws, a thirst for knowledge about nature, an
identification with the prey ("to be a deer hunter, you must learn to
think like a deer" has long been a cliche in the hook-and-bullet press,
but it nevertheless gets repeated year in and year out by outdoor writers), and a commitment to utilize the bounty in ways that honored the
wildness and uniqueness of the quarry.*

Hunters on the Firing Line

The ideal of the sportsman is now thoroughly embedded in the hunting fraternity's sense of itself. In their publications and in hunters'
public statements, hunters are described as the vanguard of conservation, true environmentalists, bound by a code of honor that respects

---

*Almost all sporting publications have a monthly food column in which recipes are featured.
More impressive, even the most plebeian rod-and-gun clubs put on one or more game dinners
each year. While not likely to feature the highly crafted terrines of duck or larded tenderloins of
venison with which the upper-middle and upper classes grace their harvest tables, the ritual of
these dinners is meant to separate game from ordinary table fare, to mark the consumption of wild
flesh as a distinct event apart from subsistence. Sportsmen do not hunt or fish "for the pot."

property, the nobility of wild animals, and the safety of others, hunters and nonhunters alike. Hunters, in this self-styled view, hunt for sport, in the transcendent sense of the word. It is not cheap thrills or base pleasure they are after. Rather, it is deep communion with nature stripped to its bare essentials. The ideal of the honorable outdoorsman has become part of a catechism that is conveyed informally, typically from father to son, and it is celebrated in rituals of the first kill, the big buck, and countless private markers of ever deepening commitment to the sporting encounter with nature.

Ron Boudin, a contractor who has been going to the Quabbin ever since his father began taking him there to fish and hike when he was eight or nine years old, epitomizes the sportsman ideal. His soft features were well tanned when I interviewed him in midsummer. He smiled when I noted that he looked like he'd been outside quite a bit. With the regional economy slumping, most contractors in the area were hustling to stay busy. Ron had been able to choose his leisure time rather than have idleness forced upon him. "I've been lucky," he commented, "to stay busy enough to be able to enjoy the out of doors without worrying about paying my bills." Ron is avid: he fishes all summer long and hunts through the fall and early winter with bow, antique black-powder muskets, as well as with modern shotguns. He is least interested in the shotgun season for deer: "Just too many people. I just don't feel relaxed or safe out there."

He prefers bow hunting, but not just for reasons of personal safety. Ron finds bow hunting compelling because it is so exacting and because it requires intimate knowledge of the woods, deer, and oneself. Such hunting also instills respect for nature.

> I don't do anything in the woods that will leave any signs other than maybe some tracks. I try not to damage any trees, so I don't build tree stands.

Ron spends hours afield before the deer season begins, scouting for signs of deer and communing with nature. Though intensely observant in the woods, hunting also offers relaxation. When I asked him what was most important about hunting, Ron was quick with an answer.

I would have to say the absence of work and phones and the routine that you have during the day. Whether you work with your hands or are an office worker or a college professor, it's time that you can get into your own head or think about nothing, which is also a nice thing to do sometimes, not having anything on your mind at all other than watching a leaf tumble from a tree.

Responsible, considerate, thoughtful—Ron could easily be held up as a model hunter. But he insisted that he wasn't unusual. As far as he was concerned, the vast majority of hunters were just as conscientious as he was. Charlotte Koski agreed. An open and animated woman in her late thirties, Charlotte lives with her husband, Tom, and their two children on a dirt road that traces the boundary of MDC land. She's a registered nurse and he is a factory worker. They have a modest house in the middle of nowhere. For them, the place is close to paradise. Charlotte recounted for me the thrill of watching a bald eagle hunting food for its young in the meadow that stretches out from their dining living room, while she was nursing her first child. The window that separated the two parents was as nothing compared to the bond she felt between them. In her view,

> being a hunter requires you to respect what you're going for. . . . That deer is not going to stand still and he's not going to walk up to you and say, "Here I am, shoot me." The hunter has to have knowledge of the environment and has to know his own limits.

Views of this sort were echoed by all the hunters I interviewed, with one exception, to whom I shall return later.

Respect, as most hunters used the word, meant more than being conscientious and thoughtful. It also meant taking pains to keep the contest between hunter and prey a fair one. For Ron Boudin, this meant disdaining gimmicks and gadgets that might diminish the contest between him and the deer he stalked. The question of a fair chase in fact disposed some hunters to take no interest in hunting the Quabbin. At one of the public hearings I attended, I overheard a hunter explaining to his buddies why he wouldn't bother applying for a permit for the hunt. "There are deer everywhere in there and they are tame as hell," he remarked. He went on to note that "any damn fool could get a deer there when they open it up." In the absence of challenge, it would not be a "hunt" in any meaningful sense of the word.

Shooting and killing, in this scheme of things, is not hunting. Focusing on the kill to define hunting is like reducing sex to orgasm. Hunting entails shooting and killing, but these are the consummation of an intricate and highly charged activity, not the activity itself. Ron was explicit on this point. Speaking of his preference for methods that let him get very close to a deer, thus establishing a kind of intimacy between himself and his quarry, he said emphatically:

> I'm a hunter. I'm not a killer. I never felt great about killing a deer or even taking an animal. I've felt successful and somewhat satisfied and I enjoyed eating the meat, but it's not a "killing thrill" that I get out of hunting.

Earlier in this century, the Spanish nobleman and social philosopher, José Ortega y Gasset, wrote a justly famous essay, *Meditations on Hunting*, which stated the matter squarely and elegantly.[4] He wrote:

> [A]s the weapon became more and more effective, man imposed more and more limitations on himself as the animal's rival in order to leave it free to practice its wily defenses, in order to avoid making the prey and the hunter excessively unequal, as if passing beyond a certain limit in that relationship might annihilate the essential character of the hunt, transforming it into pure killing and destruction. (45)

It is doubtful that most hunters, including Ron Boudin, are aware of Ortega, much less that they have read what he had to say about the hunt. And yet, they have imbibed so deeply of the rhetoric of hunting, a rhetoric framed by writers who have read Ortega, that they speak about hunting as though they had long ago memorized his essay. Ortega based part of his celebration of hunting on the necessity of the hunter becoming a part of nature, stripped bare of the advantages our superior intelligence affords. Compared to other predators, we are inept. We are slow, our vision is poor, our sense of smell rudimentary, and our hearing far from acute. Though we can bend our intelligence to the task of compensating for our physical limitations, the genuinely fair contest would seem to be tilted in favor of the wild animal. The hunter, Ortega notes, freely renounces his supremacy over nature. He continues:

Instead of doing all that he could do as man, he restrains his excessive endowments and begins to imitate Nature—that is, for pleasure he returns to Nature and re-enters it. (51)

So a hunt, in the full sense of the word, requires that hunters devise a level playing field and then behave with circumspection. Thus do hunters honor their quarry. The abundance and tameness of the deer in the Quabbin made it hard, if not impossible, for hunters to devise a level playing field. Dave Henderson, a retired public utility employee who in his retirement has become very active lobbying on behalf of sportsmen's issues, was adamant about the need to reduce the deer herd on the reservation and about the appropriateness of enlisting hunters to do the job. But he was equally certain that he had no personal desire to participate.

> I think a deer is a fine game animal. I've shot many of 'em and I have no qualms about shooting them. But I think it's an awful waste . . . to just go out and slaughter 'em. . . , from a moral standpoint. . . . I have deer hunted every year since I was in high school. I've hunted Maine, Vermont, New Hampshire, New York, Indiana, and Massachusetts. But as for going over there [to the Quabbin] I don't think there's any doubt in anybody's mind that you've got to be awfully stupid or a lousy shot not to get a deer there. . . . No challenge. I'm not interested in it.

The Koskis, who live next door to the Quabbin, also subscribed to the ideal of the fair contest, but they made explicit what most others, hunters, nonhunters, and antihunters, either left implicit or ignored altogether: true hunting not only requires self-restraint on the part of the hunter, it also requires fit animals. Charlotte Koski was telling me about the ease with which she could observe deer around her home, no matter the time of day or the season, when she added:

> The other thing we've noticed in the last couple of years, which I think is really strange, is deer around here are acting very bizarre. It's nothing to have a deer walking down the middle of the road and be totally oblivious to you. I don't know what's causing this but they are just oblivious. Others around here are reporting the same thing. Nothing phases them. They just amble along. It's very strange.

The point was clear. The Quabbin deer, because they had not been hunted and had grown so numerous and so tame, were no longer really what nature intended them to be. From the hunters' point of view, the deer were a mere shadow of what they should have been. Some of the opponents of the hunt disagreed, disputing the claim that the deer were malnourished or on the brink of a population crash. And Paul Lyons, the wildlife biologist for the MDC, felt that the deer were "still very healthy and very happy and surviving well." But this misses at least part of what the hunters were claiming. The hunters did think the deer on the Quabbin were smaller than they should have been,* but more important, they insisted that the deer at the Quabbin, small or not, had become degenerate. From the hunters' perspective, the deer were not only ruining the forest, they were ruining themselves. While hunters agreed with the MDC goal of reducing the size of the deer herd, they had their own goal: hunting would return the deer to their true, their wild, state.

Hunters, like their antagonists in the animal rights movement, idealize deer. Indeed, there is considerable overlap in the qualities attributed to deer by each side. But where the animal rights advocates speak of rights, hunters are inclined to speak of dignity, which they define in terms of the animal being able to behave as it was intended to. For deer to hang around for stale Wonder Bread or handfuls of Cheese Whizzes was disgusting to the hunters I spoke with. Deer are meant to be shy and extremely elusive, able to melt into the landscape at the merest hint of danger. They are not meant to loll about like cattle in the open.

Hunters were confident in their claim to know what deer were meant to be like. Intimate knowledge of nature, unmediated by abstracted categories of philosophy or biology, comes, in the ideal of the sportsman, from repeated, direct observation. The knowledge thus acquired has all the marks of simple truth—"I saw it with my own

---

*Based on the data gathered at the checking stations where each deer killed was taken to be legally tagged, the deer at the Quabbin were in good shape. They were not appreciably smaller than the statewide average. Whether they were "happy" or degenerate is anybody's guess.

eyes." From the little that has been written about hunting camps, it is clear that a staple of the conversations that take place, sometimes late into the night despite the early call of the morning, is detailed accounts of the behavior of wild animals the hunters have observed over the years. From the antics of chipmunks to the deadly earnest work of a coyote or bobcat, hunters share with one another a store of naturalist lore from which everyone is invited to infer the character of the species under consideration.*

This is just the sort of wisdom that Thoreau so valued, coming as it did from an ease and close familiarity with the wild. In *Walden*, he writes admiringly of the knowledge of nature possessed by the men who fished the pond as part of a complex set of subsistence activities that also included hunting.

> Early in the morning, while all things are crisp with frost, men come with fishing-reels and slender lunch, and let down their fine lines through the snowy field to take pickerel and perch; wild men, who instinctively follow other fashions and trust other authorities than their townsmen, and by their goings and comings stitch towns together in parts where else they would be ripped. They sit and eat their luncheon in stout fear-naughts on the dry oak leaves on the shore, as wise in natural lore as the citizen is in artificial. They never consulted with books, and know and can tell much less than they have done. The things which they practice are said not yet to be known.[5]

While hunters may not be scientific students of animal behavior, the best of them are systematic and thorough in their knowledge of the species they are most fond of hunting. And while they tend to talk in terms of species, their tales often feature the cleverness and elusiveness of an individual animal—the legendary buck that is tracked and stalked over several seasons to no avail, always inventing just the right ruse to elude the hunter. To regard animals in this way, to record and memori-

---

*Two recent books, one focused on New England, the other on the South, have opened up for serious consideration the culture of hunting and hunters. See Stuart A. Marks, *Southern Hunting in Black and White: Nature, History, and Ritual in a Carolina Community*, Princeton: Princeton University Press (1991); and John M. Miller, *Deer Camp: Last Light in the Northeast Kingdom*, Cambridge: M.I.T. Press (1992).

alize their cleverness and adaptability and the strength of their will to survive, is to honor them, admire them, and, yes, to "love" them.

Though the hunters would not take kindly to being called "nature lovers," they are in fact just that. They are fascinated by the drama that nature unfolds before their watchful eyes and, if my interviews are any indication, they are eager to share what they have seen and the lessons they draw. The point of sharing anecdotes (and every hunter I interviewed had his or her store of them) is to establish the claim, essential to the claim of being a "sportsman," an ethical hunter, that they are serious about nature. They are students—as Charlotte put it, "There's always something new to see in the woods." They are not marauders or thoughtless cretins intent only on killing anything that moves. Hunters claim to know things about nature that are special, available only when the senses are made acute by becoming a predator. Presuming to know what "wild" really meant, the hunters scoffed at the critics of the MDC who claimed to be protecting the wildness of the Quabbin against the predations of loggers and hunters.

Confident that they knew what nature was like, the hunters were eager to see nature restored on the reservation and they saw hunting as one of the mechanisms of the restoration. William Healy noted, in response to an attack leveled at hunting during one of the public forums, that humans have been a major predator of deer for the past ten thousand years. Charlotte, who was at that meeting, referred to Healy's comment when I interviewed her. She drew precisely the implication that Healy intended: "I feel that by prohibiting hunting [on the Quabbin] we've made it unnatural."

Even when hunters acknowledged that the Quabbin hunt would not be much of a hunt, they were willing to help restore the deer there to their natural condition, the condition that only predation can keep them in. Tom Berube, the president of the Massachusetts Council of Sportsmen, not surprisingly volunteered that hunters were willing to "be a tool of [MDC] management." The hunt, in his view, wasn't about sport at all; it was about bringing things back to normal, back to a point where the deer could be honored and respected for their real traits, not for their tameness.

This was an inspired attempt to bolster hunting's sinking image as well as a plea for a particular idealization of deer—deer as the ultimate worthy quarry. Hunting is not merely sport; and it is not simply an indulgence in nostalgia, a reenactment of a primordial activity. Hunting is environmentally and socially responsible. Hunters are "tools of management," without whom things will go plum to hell. A local outdoor writer who takes pains to urge his fellow outdoorsmen and women to be responsible and thoughtful regularly devotes his column to describing the spread of Lyme disease and of rabies as examples of what can happen when wild animal populations are not regulated by hunting and trapping.

The hunters do not think of themselves as outsiders who barge in to disrupt and distort an otherwise tranquil scene. On the contrary, they see themselves and their activity as an integral part of the nature of things. From their perspective, eliminating hunting is just as foolish and shortsighted, environmentally, as exterminating wolves. Environmentalism, for hunters, is not trying to minimize our contact with nature. It is accepting our role in nature, accepting ourselves as inevitably a part of nature, subject to its cycles and laws, and behaving accordingly. That is how hunters like to think of themselves—committed to fair play, observant and knowledgeable, socially and environmentally responsible.

Even when necessity leads some hunters to break game laws, they still can behave respectfully toward nature. Charlotte Koski spoke candidly of some of her neighbors who hunted and fished more out of necessity than for recreation. Like subsistence hunters everywhere, they did not feel particularly encumbered by the laws regulating bag limits and seasons.

> I know some people who environmentalists would probably consider very bad but I think I see more respect in them for the animals in the woods than I do in most people who come out here [from the city] and build and have no idea about what nature is. They may hunt out of season but it's not because they are greedy, they need the food. And I see them taking more care and concern not to harm animals. . . . They may not be very socially respectable but they certainly have respect for wild things.

Living close to the land and to nature, whether of necessity or by choice, is presumed to bring out a sense of reverence for nature that cannot be replicated by urban-bred environmentalists who relate to nature abstractly.

If this were all there was to the matter, hunting would not be as beleaguered as it is and hunters would not be as defensive as they are. The emphasis on sportsmanship, on hunting for the experience of being a part of nature, not for the rude business of filling the pot, on social and environmental responsibility, masks another side of hunting, the side that troubled the Puritan guardians of order and rectitude. Hunting also can bring out the presocial in humans, that part of humans that is antagonistic to rules and regulations and the to self-restraint that makes civilization both possible and fragile.

There is something about hunting that invites people to tempt fate as well as to defy social convention. People go hunting for many and complex reasons, but among them, as Ron Boudin's earlier observation makes clear, is the desire to get away from the normal routines and constraints of the work-a-day world. Some of these constraints involve etiquette and impulse control, and require accepting frustrations and learning to live within the rules. Hunting, by contrast, makes it possible to devise your own rules and to break them if the impulse strikes. Being in the woods means that you are not easily scrutinized; you can do things there that you cannot get away with in town. These run from the trivial—relieving yourself "when nature calls"—to the potentially serious—shooting for the hell of it.

In addition to tales of adventures with animals, hunters have their stock of stories about such transgressions large and small. There is the bird taken out of season—because the dog had worked it perfectly and the gunner felt he had to reward the dog by shooting the bird. Or the shot taken too near a house—but in a safe direction—that "must of scared hell out of folks, ha, ha." Fences get broken, livestock, pets, and outbuildings get shot at, and otherwise ordinary people become rude and arrogant.

The official line on this sort of behavior from the hunting fraternity is that it is the work of a few "slob hunters," men who are not true sportsmen. They violate game laws, endanger the public safety, enrage landowners by their rudeness and disrespect for property, and give the whole of hunting a bad reputation. No doubt, there is truth to this. A few "bad apples" can cause the whole barrel to rot. But things are more complicated. While only a very small handful of hunters may be honest-to-god unrepentant "slobs," people who are just plain stupid and boorish, almost everyone who hunts or fishes occasionally behaves like a "slob." Many of the violations of the sportsman ideal occur in a hunter's adolescence and young adulthood, when manhood has yet to be firmly established and the need to dominate, to conquer, might arguably be at its peak.* But there is a bit of the boy in most men, if behavior at college reunions is any indication.

One of the hunters I interviewed, a man who lives adjacent to the Quabbin Reservation, first drew my attention when, at one of the public hearings, he rose, declared himself in favor of the hunt, and then, to establish his lack of self-interest, noted that he had been convicted of poaching on the reservation and thus was forbidden from even setting foot on MDC land. In our interview, he explained that he was innocent of the charge but could have proved that fact only by fingering a friend—now an ex-friend—who had, indeed, killed a deer. This was not a question of a man needing to feed his family. And since, as we have noted several times, the deer on the reservation are nearly as tame as cattle, there could have been no sporting challenge involved. The only contest was between the poacher and the authorities.†

The thrill of getting away with something, even if the long-run consequences of doing so are injurious to yourself, is clearly a force shaping hunter behavior afield. The promulgation of the sportsman ideal and its continuous promotion by outdoor writers, equipment manufacturers and retailers, state fish and game bureaus, and the legion

*A number of studies have shown that hunting patterns change over the life course. In adolescence, hunters tend to be preoccupied with firepower and define their satisfaction in terms of how many animals they kill. Over time, hunters shift their priorities. Getting away becomes as important as bagging an animal. Emphasis shifts to the more ethereal and aesthetic. For an example of this research, see Bob Jackson and Bob Norton, "Hunting as a Social Experience," *Deer & Deer Hunting*, (November/December 1987) 38–51.

†This is not to say that poaching cannot be sporting. The skills of the poacher might indeed have to surpass those of the legal hunter. The latter "only" has to outwit his quarry; the former has to outwit his quarry and the law. See Ragnar Benson, *Survival Poaching*, Boulder, CO: Paladin Press (1980), for an unabashed celebration of what the general hunting fraternity condemns as "slob" behavior.

of rod-and-gun clubs across the nation has undoubtedly done an enormous amount to reduce the "slob factor," but hunter misbehavior is still a nagging and serious problem. The excitement of the hunt, coupled with the sense of freedom that comes from being in the woods, produces a context in which the temptations to misdeeds and foolishness are many and momentary lapses in judgment are always a possibility. Again, this is not a matter of "bad people." Hunters are on vacation when they hunt, and as with vacationers generally, their inhibitions are lowered. After all, resort behavior, even among the most refined and cultivated, is not widely regarded for restraint and good taste.* Hunters, even the best of them, may succumb to such temptations at least once in a while. This makes hunters as suspect now as they were in Plymouth Colony nearly four hundred years ago.

As I noted earlier, several of the opponents of the hunt had themselves been hunters and one still actively hunted. Unlike their animal rights allies, these former hunters objected less to hunting in the abstract than to the behavior of hunters, especially what they saw as hunters' cavalier disregard for the environment and their thoughtlessness. Ray Asselin ruefully noted his own past misbehavior as evidence of the excesses of hunters generally.

> I used to be a hunter and that's all I lived for. I was pretty bad when I was a kid as far as killing animals went. I would kill anything and everything. . . . I don't have a problem with a really good sportsman who wants to go out and responsibly spend a day in the woods and if he shoots some game, fine. But I have seen so many slobs out there, and idiots. I was kind of irresponsible myself so I know what it's about. I don't know how you could weed out the idiots so I suppose you just stop it all.

Bill Granby had probably had more access to high-quality hunting range than anyone I encountered in this project. As an officer in the

---

*John Mitchell, long a writer on the staff of *Audubon Magazine*, explores both the admirable and the seamier sides of hunting in his fine book, *The Hunt*, New York: Knopf (1980). He notes, for example, that deer season turns some small towns in northern Michigan, a state where deer hunting is very popular, into temporary watering holes and brothels, a transformation different perhaps only in coarseness from the one that the summer season brings to coastal retreats for the upper classes.

Armed Forces, he was stationed in or near prime spots for big and small game on several continents over the course of his career. He hunted avidly until he grew sick of guns and killing in general, largely as a result of what he saw in Vietnam. Now he regards most hunters as "very shortsighted people."

> They're not interested in the environment, they're interested in blowing the hell out of something. . . . I wasn't one of those morons that go out with spotlights and a rifle and beer. Sure, some do it right but most do it wrong. They're people who don't quite grow up.

Greg Kuznets, now in his seventies, still enjoys hunting ("where can you find an activity where you can loaf for a month without somebody calling you a lazy so-and-so?") and knows many conscientious hunters. But as a former employee of the Division of Fisheries and Wildlife who helped stock trout and pheasants, he has witnessed some disgusting behavior by "sportsmen."

> I've seen plenty of abuses. People think "It's in the country, to hell with it." And then you have the people who think the Lord put it there for them and they're going to use it, abuse it, do whatever they want with it.

Hunters have absorbed the sportsman ideal, even if they can not fully make themselves over in its image. They can entertain each other with examples of how the ideal has been violated because they share an ideology that rejects such behavior. The lapses are just that, lapses: fleeting exceptions to the rule that provide comic relief, often at the expense of the timid, oversocialized city slicker and the much despised "Bambi lover." But the nonhunting public is, at best, only dimly aware of the sportsman ideal. What they see are unkempt armed men and what they hear are stories of horrible accidents and inexcusable violations of good sense and common decency.

The gulf between hunters and nonhunters is made wider by virtue of the differences between the two populations. Hunters are overwhelmingly male and are disproportionately drawn from the small town yeomanry—men who work with their hands or who service those who do. The hunting fraternity does also include a slice of the nation's white-collar population—salesmen, professionals and businessmen—but the center of gravity is clearly working class.* This means that hunters' language tends to be coarse to middle-class ears, that traditional assumptions about gender and sexuality are more deeply held, and that there is little patience for squeamishness or sentimentality (unless it is directed toward dogs, nostalgic evocations of the good ol' days, or favored hunting grounds). "Hard living" is more common, and its signs—scars, tattoos, bad teeth, beer bellies—more evident. With blaze-orange garments calling urgent attention to themselves, hunters *do* look like roughnecks and renegades. Add the conspicuousness of a firearm and it is not hard to see how nonhunters could form a vivid and uncharitable impression of hunters.

Demographic trends have also contributed to the negative reaction to hunters. Before World War II, the inhabitants of small town and rural America did not rub elbows with residents of our larger cities very frequently. The cultures of the two settings were distinct. Hunters hunted in areas populated by people more or less like themselves; people, that is, who knew one another's values and life styles. It was the city folk, almost always of higher status than the country dwellers, who did the foolish things in the country, whether they were merely sightseeing or camping, or hunting and fishing.

Now, however, the city has come to the country in the form of suburbs and exurbia. Beginning in the late 1940s, America's cities have poured forth a constant stream of middle- and upper middle-class people seeking solitude and the tranquillity that the hinterlands seem to promise. Here, they run smack into practices like hunting that they do not understand and for which they have little sympathy. Men with guns have no place in their idea of a rural idyll. The result is inevitable: land gets posted. Combined with development, posting has dramat-

*The portrait drawn here is based on studies carried out periodically by the U.S. Forest Service and the U.S. Department of Fisheries and Wildlife. In its 1980 survey of users of the out-of-doors, the Fish and Wildlife Service found that 69 percent of all hunters lived in small towns and rural areas, 92 percent were male, and only 8 percent had annual incomes exceeding $40,000. By comparison, nearly 40 percent of all American families enjoyed incomes of over $40,000 in the same year and only 26 percent of the general population lived in small towns and rural areas. U.S Department of Interior, Fish and Wildlife Service and U.S. Department of Commerce, Bureau of the Census, *1980 National Survey of Fishing, Hunting, and Wildlife-Associated Recreation*, U.S. Government Printing Office, Washington, DC (1982).

ically reduced huntable land in densely settled states like Massachusetts. More rural states, like Vermont, have been transformed by the growth of ski areas, and the condo villages and fancy year-round second homes of the urban elite that grow up in the vicinity of the slopes, all of which occasion a flurry of "No Trespassing" signs around the expansive perimeters of these upscale enclaves.

Development and posting have forced hunters to travel farther and farther to find suitable areas in which to hunt. In turn, this has meant that less and less hunting takes place within a moral economy in which norms are enforced by virtue of close familiarity. Now, hunters are anonymous. They are not your neighbors or cousins or the guy that repairs your car or sells you plumbing supplies. There is no particular reason to trust them and, on the hunters' part, there is no strong bond that stimulates conscience or solicitude for others. However much hunting, by itself, may encourage uncivil behavior, when hunters gain anonymity, inhibitions on impulse grow even weaker.

Groups of men roam the back roads, a few (which is all it takes) openly enjoying beer—and thoughtlessly throwing the cans out the window—and leering at the locals, looking for likely spots to hunt with less apparent intensity than places to relieve themselves. Unfamiliar with the area, even the well behaved of these outsiders make mistakes: they get turned around and wind up in someone's backyard instead of the back forty. Thinking they are in deep woods, they wander close to roads or houses or other areas in which those more familiar with the local scene would not dream of hunting. Posted signs, the consequence of locals taking umbrage at the "invasion," get shot up and torn down, further adding to the image of the hunter as rogue. Cut off from the local moral community, hunters, whether they meet the sportsman ideal or not, are seen as a threat to peace and safety. People who hunt themselves or who are accustomed to hunting can feel beleaguered and put upon, even when none of the hunters in a given area or on a given day does anything wrong, discourteous, or foolish.* The sportsman

---

*Even the most ardent hunters and supporters of the Quabbin deer hunt did not want to see the Quabbin opened up to unrestricted hunting, precisely because they did not relish the prospect of hordes of hunters sweeping through the reservation. Charlotte Koski, reflecting on what an unrestricted hunt would mean for her, living a stone's throw from the reservation, said bluntly "I don't think I would enjoy the hunting population that would come into this area." Her husband,

ideal may make hunters feel responsible and thus entitled to pursue their sport; it does not so easily persuade the middle-class nonhunters of the virtue of hunters.

Indeed, the MDC staff shared the general middle-class disdain for hunters. Part of this antipathy was rooted in their experience with "sportsmen" who were allowed to fish the reservoir. The MDC has strict rules against picnicking and drinking, in part to control litter and in part to minimize urination and defecation on the watershed or in the reservoir itself. But routine patrols turn up all manner of debris and other evidence of abuse of the privilege of being allowed to fish on MDC property. This experience has been dispiriting, to say the least. It is not so much that the purity of the reservoir has been endangered—though this possibility cannot be ignored. The refuse is an assault on the *idea* that the Quabbin is special and that it should be treated reverently.

Sportsmen have been in an adversarial relationship with the MDC for a long time over access to the reservoir for fishing. From the agency's perspective, organized sportsmen use conservation as a rhetorical cover. Down deep they are interested in one thing and one thing only: consumptive use of fish and game. This single-mindedness, coupled with bluster and rudeness, offends many in the MDC faction. Thom Kyker-Snowman was convinced of the need to drastically reduce the deer herd, and he eagerly looked forward to the "recovery of the forest" as a consequence of the hunt. "But," he confided,

> I don't look forward to managing a sport hunt because I've certainly encountered my share of difficult people in the hunting community. Especially the deer hunting community.

Clifton Read has tried, in his role as interpretative naturalist for the MDC, to persuade sportsmen that the agency was not interested in expanding recreational hunting opportunities. The point was to get the deer herd down to manageable size as quickly and efficiently as possible. The response?

Tom, nodded his agreement and added that that's why they supported the MDC plan that called for tightly controlled numbers of hunters and a high degree of monitoring of the hunt itself. Ron Boudin, who lives miles away, was equally disturbed by what it would be like were the Quabbin simply thrown open to hunting. "A God-awful mess" was how he described it.

They wouldn't hear it. They were [only interested in the] great opportunity for hunting. The whole thing about this being a controlled hunt they didn't hear, didn't understand it.

Clif went on to report how the opponents of the hunt characterized hunters—"people who shoot at everything that moves"—and while he knew full well that this represented classic stereotyping, he concluded:

Unfortunately there are enough hunters out there [who are like that]. Even though they may be a small percentage, they are a concern. They may not really care about what we are trying to achieve and if they act irresponsibly, they could jeopardize the whole effort.

This dislike of hunters (as distinct from a dislike of hunting) is one of the reasons it took the MDC several years to reach the conclusion that the deer had to be hunted. It is fair to say, though the critics remained doubtful on this score, that the MDC looked carefully at every plausible alternative to hunting. The agency explored nonlethal alternatives not just because they were committed to thoroughness, nor because they were eager to avoid controversy: they did not relish the prospect of hunters on the reservation. Their experience with fishermen and their negative feelings about hunters made them very reluctant to move in the direction of a hunt. Robert O'Connor summed up the MDC staff's sense of things this way:

Everybody here knew what a big headache it was going to be to do something about the deer so we were just hoping that it would just go away, or maybe we could just continue to dabble with these fences and that would be good enough. That's why it seems that we had contradictory policies.

As the alternatives to shooting the deer were ruled out, because they seemed either unworkable or too expensive, the MDC was forced to "bite the bullet." Some sort of hunt would be necessary if the watershed was to remain sound and resilient. It was at this point that the divisions that we have been tracing among the supporters of the MDC's Quabbin program began to emerge. As supporters and friends moved into opposition (and sometimes personally rancorous animosity) over the issue of killing deer, the MDC and those who agreed with their conclusions found themselves with a strange and unsettling ally: sportsmen.

Hunters: "Tools of Management"

Given the low opinion in which hunters were held, it is not surprising that the MDC was not happy with the prospect of reducing the deer herd by hunting. If lethal means were necessary, they would have preferred sharpshooters to sport hunters for the job. Sharpshooters were preferable for several reasons. First, they could be tightly regulated and supervised. Moreover, being "mercenaries," they would be more likely to comport themselves as professionals or specialists than would your average sport hunter—that is, they would accept supervision more willingly. Second, it was assumed that sharpshooters would be more businesslike, that they would be interested in getting the job done as quickly and with as little fuss and muss as possible. Finally, sharpshooters would not appreciably increase the number of people who might come to regard the Quabbin as theirs and who would thus form a lobby, much like the fishermen had, that would bedevil the MDC for years to come. Sharpshooters, in a sense, would be the analogue to the loggers with whom the MDC contracted to cut the trees that the agency's foresters marked for cutting.

Thom Kyker-Snowman voiced the general sentiment, the "druthers," of the MDC as well as the preference expressed by the nonhunters who supported the proposed hunt.

> I guess if it were up to me . . . the best thing to do would be to bait 'em [the deer], spotlight 'em, and get a sharpshooter and put 'em down and distribute . . . the meat to the homeless shelters. . . . I think that's a great idea and I think that's what should happen.

The MDC had, in fact, contacted food banks and correctional facilities to see if they could use the venison coming from a sharpshooter program. But the real issue was hardly getting food to the needy. The MDC wanted to reduce the deer in the same *spirit* as they harvested trees— methodically, carefully, scientifically. Paul Lyons put his finger on the heart of the matter. "We [should] be doing what we feel is needed here in the most efficient way possible, without adding this recreation or fun or sport aspect to it."

Sport hunting was simply inappropriate to the Quabbin, even in the eyes of those who had come to accept the necessity of killing deer. Elisa

Campbell had been reflexively antihunting, not least because, in her long experience as a Sierra Club activist, she had found that sportsmen could rarely be counted on to support the kind of environmental efforts she felt were urgently needed. She wanted the deer herd reduced as quickly as possible without opening the reservation to greater public access. "Use bait, lights, and aim for does," was her prescription.

There was even the intimation, before the option was finally dropped, that some in the animal rights contingent would go along with a sharpshooter program. Apparently, though, private flexibility did not survive the glare of public negotiations, and when it came down to agreeing to go ahead, the animal rights lobby refused to endorse a sharpshooter program. Thus, as things moved toward a conclusion, support for sharpshooters narrowed decisively. Three very different sorts of problems with sharpshooters rapidly surfaced as final recommendations began to take shape. First of all, no one could say for sure who was and who was not a "sharpshooter." The Armed Forces and the Boy Scouts use such a designation to acknowledge a certain competency in hitting targets, but the applicability of that kind of standard to shooting in the woods at live targets was questionable.

Related to this ambiguity was the fact that it takes more than an accucrate shot to kill a deer. Even "tame" deer wise up quickly. Experience in other areas with an overpopulation of deer have shown, for example, that baiting (putting food or salt licks out to attract deer) stops bringing deer in almost immediately once the shooting starts. After an initial flurry, during which unsuspecting deer can be shot rather easily, killing deer becomes a matter of *hunting* as opposed to shooting. Successful hunting requires a knowledge of terrain and the habits of deer, and a willingness to put up with inclement weather, as well as the ability to shoot accurately. Sharpshooters, assuming they could be identified, are specialists only in shooting.

It is easy to understand how people unfamiliar with guns and hunting could overlook such details as these. In a culture filled with images of SWAT teams and "surgical strikes," and much inflated talk of high-tech weapons, killing deer might seem like a simple matter to the uninitiated. It is not. But even if it were a simple matter from a logistical standpoint, there was one final problem with the sharpshooter option. The deer on the reservation were not under the jurisdiction of the Metropolitan District Commission. By law, the deer are under the aegis of the Division of Fisheries and Wildlife, and they can be killed only in a manner and at times agreed to by the division. The division was dead set against a sharpshooter program.

Steve Williams was head of the division's deer program. Young and plainspoken, he summarized the problems with sharpshooters, including the uncertainty of establishing credentials, and added a few notes of reality of his own.

> Think of the cost involved. Those who think that you're going to get a volunteer group of so-called sharpshooters to work year-around, and that's what you have to do to have hope that it would work, that they are going to do that on a volunteer basis, are sadly mistaken. It's very expensive in terms of man-hours, which translate in money, man-hours per deer. And you've got to cover . . . fourteen square miles.

Ironically, as we began our conversation about the limitations of sharpshooters, Williams was interrupted by a phone call. A deer had been spotted in a small enclosed area adjacent to a major freeway interchange. Shooting the deer was risky because the area was surrounded by homes, a hospital, and office buildings. Tranquilizing was problematic because the deer would be likely to bound onto the freeway before the drug took effect. While his staff pondered what course of action to follow, we resumed our conversation.*

> We view white-tailed deer as a public resource, as an asset. And we feel very strongly that as an asset, as a resource, they should be treated as such. And that means harvesting and consumption. . . . We don't make any apologies for that. And to restrict that activity or to restrict the public from having access to a public resource, we do have a problem with that.

In defending the division's position, Williams pointed to past experiences with so-called sharpshooters. One example was from Mas-

---

*Episodes like this grow steadily more common, an indication of what may turn out to be an embarrassingly successful effort to improve wildlife habitat and reproduction rates. The *National Geographic* recently ran a feature on this problem. Wildlife populations are growing throughout the East and are clashing with a wide range of human activities, from golf to commuting. See James Conaway, "Eastern Wildlife: Bittersweet Success," *National Geographic*, Feb. 1992, 66–89. We shall return to this matter briefly in chapter 6.

sachusetts. On the coast, a small reserve, Crane's Beach, had been set aside years ago. Hunting was forbidden and the deer multiplied to the point where both the reserve and the deer were headed in a dramatic downward cycle. To make matters worse, Lyme disease became endemic in the deer, and the tick-borne disease swept through the several hundred families with homes adjacent to the reserve. It is estimated that over one-third of all the families abutting the reserve now have at least one family member afflicted by the disease.

The trustees of Crane's Beach faced the wrath of the animal rights activists when they proposed bringing the deer herd down to sustainable numbers. As a compromise, all parties agreed to allow sharpshooters in to do the dirty work. It did not work well, because after the first shots rang out the deer refused to cooperate, becoming the elusive creatures hunters so admire. The next year the area was opened to a controlled hunt that proceeded without a hitch and resulted in many more deer being taken. Williams concluded from this and similar experiences across the country that: "The people who have been hunting deer in this state are the best qualified to hunt deer."

Considering the fascination with guns in our culture, it should not be hard to find people willing to sally forth and shoot deer more or less as target practice. In fact, volunteers are few. Part of the reason is that after a century or more of idealizing the white-tailed deer, everyone has a hard time thinking of the animals as pests or vermin. Hunters are quick to ridicule the idealizations of the antihunters, whom they disparage as "Bambi lovers." But hunters, as we have seen, also idealize the deer. The white-tail is one of the most challenging of all game animals. Its grace, speed, and wariness combine to make hunters admire the animal. Killing a deer is an accomplishment of which they are proud. Only a handful of hunters, comparatively, are successful more than a few times in their lives. The thought of sharpshooters slaughtering deer is a violation of deeply held values, a breach of the compact that hunters have implicitly made between themselves and the deer they stalk.

Hunters may also resent sharpshooters because they displace hunters. More important, though, hunters object to sharpshooters because they have no respect for their targets. They are strictly gun oriented:

gunslingers, not hunters. Hunters pride themselves on their marks-manship, not as an end in itself but as a means to the end of honorably killing the quarry. Every hunter's aim is for a clean, quick kill. Sharp-shooters' motives are different: they are simply killers. The deer deserve better.

In any event, the Division of Fisheries and Wildlife had a strong case against the effectiveness of sharpshooters. Some in the MDC faction accused the division of promoting recreational hunting to defend its own constituents—the people who buy hunting licenses—and budget. They feared that the division saw the deer problem at the Quabbin as a chance to expand hunting opportunities in a state that was steadily moving toward restricting hunter access to land. They worried as intensely as the MDC's critics that once the division got control of the hunt, the hunting lobby would be irresistible, and hunting would become a routine event on the reservation.

There is little doubt that the division was more powerful than the MDC, and it may have been more adept politically. Moreover, the law was clear: wildlife in the state are subject to the division's rules and regulations. Landowners can regulate access to their land by hunters, but they cannot regulate the treatment of wildlife that may reside on or pass over their land. Period. The MDC had little choice, finally. Hunters would have to be enlisted in the effort to control the deer. Hunters would have to become, in Tom Berube's words, "tools of management."

The Division of Fisheries and Wildlife and the MDC diverged on other issues besides sharpshooters. The MDC was interested in reducing the deer herd as quickly as possible. The best way to do this would be to concentrate on reducing the number of does. Since deer are polygamous (one buck is capable of impregnating many does in a mating season), even a drastic reduction in males would have little effect on reproduction rates as long as the numbers of females remained high. In their efforts to maintain a large herd in order to please hunters, game managers historically have tightly regulated the taking of females. Their goal is to tilt the annual kill heavily toward bucks, leaving plenty of does to produce a bountiful supply for the next year's hunting.

This management strategy also plays well with the sportsmen. Bucks are warier and generally more elusive than does. That is one reason why such importance is placed on getting the proverbial "big buck." Bagging a large old animal requires the full range of skills, from woodcraft to physical endurance, that the hunter at his best possesses. Such animals are trophies not just because they are scarce but because killing one demands unusual ability. Given the choice, any hunter—as opposed to someone interested in filling the pot—would choose a buck, the bigger the better.*

For its part, the Division of Fisheries and Wildlife insisted that hunters be restricted, as they are in general, to taking only one female (they can take two bucks or one deer of each sex, if the hunter has received a so-called doe permit which the division awards by lottery to those who apply each year). The MDC, left to its own on the matter, would rather have stipulated that hunters take two does, perhaps then allowing those who had succeeded, as an added incentive, to hunt for a buck as well. But the division was adamant. The statewide bag limits would remain in effect for the Quabbin hunt: only one doe per hunter.

The MDC faction was clearly upset with this. If hunters were to become "tools of management," the MDC faction thought of them as blunt instruments at best. They preferred management that was more refined, more subtle. Moreover, the MDC faction worried lest the hunters would show a strong bias toward shooting bucks, thereby undermining the goal of permanently reducing the deer herd. No matter how successful the hunters might be, the tilt toward bucks would ensure steady reproduction of the herd. This would mean that the Quabbin would have to be open to hunting forever. This prospect, added to the general suspicion of the motives of the division and the even greater misgivings about hunters, made the MDC faction anything but enthusiastic about the turn away from sharpshooters.

Indeed, when the sharpshooter option was dropped, the division between those who still supported the MDC and those who had moved into opposition grew deeper and more acrimonious. People remaining

---

*Antihunters read this as a script for machismo—one male dominating another male. The symbolism cannot be denied, and there is plenty of lore regarding, for example, the traditions governing the disposition of the buck's genitalia to give credence to the claim. My point is that more is at work in this than a psychosexual dynamic. If does were equally elusive and the challenge of killing one the same as for a buck, the preferences of hunters might be different.

in the MDC faction who had long opposed what they saw as the narrow and self-serving environmentalism of fish and wildlife agencies were regarded as having betrayed the cause. Local branches of some organizations were severely split over this issue, and many personal friendships, some of quite long standing, were ended. Alexis McNair, who is forceful and feisty enough to have enjoyed many a good squabble, did not reflect on this particular episode with much pleasure.

> It tore the group [a local chapter of a prominent national environmental organization] right down the middle like an old piece of paper. . . . I knew it wasn't going to be fun, but I had no idea how un-fun it was going to be. This was probably the least fun thing I've done in years.

Elisa Campbell said, mirthfully, that her associates on the other side of the issue must have thought that she had been drinking "strange beverages" to have endorsed the killing of deer, especially the killing of deer by sport hunters.

However willing and eager the sportsmen were to assist the MDC in its management of the Quabbin, the actual managers and their most ardent supporters, past and present, were not happy with their newfound associates. The hunters, in a sense, were hoist on their own petard. Having deeply imbibed of the ethos of the sportsman who hunts for personal challenge and aesthetic experience, and has an elaborate ideology of restraint, of treating wild animals with reverence, hunters were not the logical choice for the sort of matter-of-fact management that the MDC would have liked. To be effective on the MDC's terms, hunters would have to suspend their customary predilections, for example by shooting does rather than bucks.

This could be easier said than done. Since the deer in the Quabbin were tame as well as numerous, the only way that hunting there could be rendered remotely sporting would be by seeking large bucks. This would mean passing up many shots at does. The MDC and the Division of Fisheries and Wildlife attempted to reduce this possibility by dividing the nine-day hunting season into three three-day segments. Past experience had taught that deer harvests are highest on the first and last days of the season. Hunter success rises on the last day because hunters who have waited for the opportunity to get a large buck will, on the last day, settle for anything. Since these hunters tend to be the most avid and adept, their success rate is high even though all the easy pickings have long since been hung in meat lockers. By having three mini-seasons of three days each, the Division of Fisheries and Wildlife hoped to create three "last day" situations, thus minimizing the likelihood that large numbers of hunters would hold back, waiting for the buck of a lifetime.

Aside from this alteration in the normal nine-day shotgun season and a minor alteration in the legal hunting hours, the division insisted that the customary statewide game laws apply to the Quabbin hunt. This did not please the MDC and its nonhunting supporters. The MDC, as we have seen, worried that hunters would treat the Quabbin hunt like other public hunting, that is, as a recreational pastime. The MDC wanted *managers*, not sportsmen, loose in the reservation, people who would approach the hunt with the idea of getting a job done, not with the idea of satisfying primal urges or embodying some abstract notion of a sporting ethos. They wanted everyone to understand that they had no intention of allowing the reservation to become another recreational asset of the state.

Having lost the arguments with the Division of Fisheries and Wildlife over sharpshooters and bag limits, the MDC was left with the task of impressing upon the hunters selected that they were there to help manage the reservation, not to have a good time. If, in the process of doing their job, hunters also had fun, so be it, so long as everyone understood that this was not sport hunting.

To drive this point home, the MDC decided to require all hunters selected for the Quabbin hunt to attend an orientation session before the actual hunt commenced. In addition to providing basic information about the area of the reservation to which each hunter was assigned, the sessions emphasized repeatedly that the hunt was being staged to bring the size of the deer herd down to acceptable levels as quickly as possible and that the hunters were being enlisted in a carefully designed program of environmental management. The MDC also outlined a series of rules and regulations that they hoped would make the Quabbin hunt unmistakably different from general sport hunting. The goal was to kill as many deer as the game laws permitted.

The distinction between hunting for management and hunting for recreation is clear on paper but by no means easy to sustain in a population long imbued with the rhetoric and symbolism of sport hunting. As hunters gathered for the orientation sessions (ten were held in all to accommodate the one thousand hunters selected from among the nearly ten thousand who had applied), they invariably speculated on the number and size of the bucks on the reservation. And the MDC staff, despite their own ambivalence about allowing hunting and hunters on the reservation, could not refrain from making tantalizing comments about the "huge racks" carried by some of the bucks. Such comments sent an excited hum coursing through the otherwise quiet and attentive assembly. Clearly, the fascination with impressive racks of antlers is not restricted to hunters.

Hunters may have been the best "tools of management" available, given the limitations on sharpshooters, but there are many points at which the ideology of hunting and the ideology of management collide. To be tools of management meant that the Quabbin hunters had to become preoccupied with the kill, something they had come to believe was almost an afterthought, not the point of the whole endeavor. Many were uncomfortable with the role of "deer killer" as opposed to "deer hunter." To add to their discomfort, the MDC and its newly acquired managerial assistants were confronted with opposition made more implacable and vociferous precisely because sport hunters were to be involved. The MDC faction, palpably uneasy with the shotgun marriage between themselves and the state's hunters, did not fully appreciate the dilemma hunters felt.

Death or Pleasure: One Man's Meat, Another Man's Poison

There is no small irony in the fact that to make hunting honorable in the nineteenth and early twentieth centuries, the emphasis had to be on sport, not meat. Now, the emphasis on sport has become central to the indictment leveled against hunters. The placards that animal rights protesters of the proposed hunt carried outside the public hearings excoriated those who kill for pleasure. Dismissing entirely the deeper sense in which hunters speak of sport or pleasure, the opponents of the hunt saw only the blood-drenched hands of sadists bent on mayhem and murder in order to assert a vulgar, brutal dominance. Bill Granby's image of hunters intent on "blowing hell out of everything" was widely shared. Lenore James and her husband, Steve, objected to hunting because, in their words, it embodied a separation between hunter and animal, an absence of "togetherness."

> Yeah, togetherness. Hunting is a separation, it's the old biblical approach that the world was given to us to dominate. . . . It gives the hunter a feeling of dominance.

Martin Dodge, the man who had devoted so much of his life to setting up preserves, referred disparagingly to the sportsman's ethos of loving nature and offered this observation:

> At least [around here] the people who hunt are literally rednecks. It seems to be something that attracts the more violent people. And many of them are the same people who are out there with their four-wheel drives, tearing up the roads and the woods and damaging the environment.

The idea that hunters—beer-guzzling drivers of gas-guzzling pick-up trucks—could be enlisted in a delicate environmental management effort was clearly a proposition that only hunters and the Division of Fisheries and Wildlife could swallow. The rest, critics and proponents of the hunt alike, albeit with different intensities of feeling, found the idea preposterous or at least unsettling.

This near-total dismay put the hunters squarely in a double-bind. Their social background made their claim to be concerned about the environment hard for the middle-class environmentalists who formed the bulk of both camps to accept. When they insisted that they were sportsmen, committed to ethical conduct and deep respect for wildlife, their critics lambasted them for murdering innocent creatures for transient pleasure—"for no good reason at all, just antlers," as animal rights activist Lorraine Near put it. And when Tom Berube pointed out that part of the ethic is to eat what you kill, he had to confront placards proclaiming "Meat is Murder."

Understandably, hunters responded with exasperation to the box they found themselves in. For them, the matter was clear and straight-forward: there were too many deer at the Quabbin and only hunters, warts and all, were qualified and competent to do something about the matter. They were perhaps not sophisticated or refined in their manner, but they were not about to yield on the question of their indispensabil-ity. As Tom Berube, chief spokesman for the state's hunters, said, "There's a job to do and we're the ones to do it. Period!"

The MDC faction had to agree, as much as it pained some of them. As they saw things, there most definitely were too many deer, and there was an urgent need for decisive action. However much they may have wished to press the sharpshooter option, they knew when they were outmaneuvered. To continue to wrangle over sharpshooters would simply delay the start of the deer reduction effort and subject the watershed to that much greater damage. In the end, the sense of the gravity of the situation overwhelmed their low opinion of the hunting fraternity.

As for the opponents of hunting, they left no ground on which reconciliation or even mutual indifference might be possible. The pro-ponents of animal rights and the hunters are unintelligible to one another. Hunters cannot fathom how people who profess love for ani-mals can sit idly by knowing that deer are going to be starving to death. Animal rights advocates cannot, for their part, comprehend how hunt-ers can claim to love the animals they so eagerly seek to kill.

Death appears to be at the center of the chasm separating the two camps. This is, though, more appearance than substance. Within min-utes of serious conversation in the interviews I conducted, animal rights sympathizers willingly acknowledged that death is inevitable and that it is rarely welcomed or "easy." Serious adults, however idealis-tic, do not imagine nature as kind, merciful, or gentle. Starvation is not a pleasant form of exit; nor is being dragged down by a coyote. From the animal's point of view, it is hard to imagine any preferential hier-archy as between the various "natural" forms of demise and the end that comes by being shot by gun or arrow. Death will come, somehow, and it won't be lovely.

Hunters argue that the death they impose is at least as humane as anything nature delivers up, if not more so. The ideal of the "clean kill" is key to this view. Hunting's critics respond, alleging that substantial wounding is routine, a fact that hunters vigorously dispute.* Whatever the truth might be, the lives of wild animals, even without hunting, are far from pain-free. Legs get broken fleeing predators; gashes are opened up in encounters with old strands of barbed wire or collisions with broken tree limbs; and so on. So it is not clear that crippling from a poorly aimed arrow or slug adds appreciably to the hazards and levels of discomfort endured by deer.

This is no small quarrel. Following Jeremy Bentham's lead, Peter Singer, the contemporary philosopher most responsible for the current upsurge of animal rights activity, argues that animals have moral claims on us by virtue of their capacity to feel pain. Pain is presumed bad, since it is intuitively obvious that every creature seeks to avoid or, if unsuccessful at avoidance, to minimize it. We therefore should do what we can to reduce needless pain and its attendant suffering. What is most odious about hunting is that it imposes *gratuitous* pain on animals, pain that is unbidden, unrelated to anything that the animal must do or endure as a result of being what it is. The coyote has no choice but to hunt for deer, and the deer have no choice but to endure the coyote's predations—this is what nature has decreed. Once we humans cease to be directly dependent upon nature, once we become able to supply ourselves with nourishment from crops and clothing from fabric, we are no longer compelled to add to the pain and injury endured by

*The issue is so highly charged that otherwise ardent promoters of research on hunters and hunting are loathe to inquire systematically about crippling losses (animals wounded but not retrieved). As a result, no one has the slightest idea what the facts really are. Numbers are bandied about with great conviction and almost no authority. Needless to say, even if there were a firm resolve to find out the truth of the matter, it would not be easy to do. Because the ideal of the clean kill is one of the foundations of the sporting ethos, there is no reason to expect hunters to be eager to admit to having been sloppy or negligent in selecting their shots. As a hunter myself, I can attest to the deep and disturbing disquiet that descends when a crippled animal is not recovered. The thought of it suffering as a result of my action weighs heavily. But such incidents occur, in my experience, "occasionally." I am not an avid deer hunter, though I do hunt for deer and have been successful twice, but I am an avid bird hunter. My crippling losses average less than one bird per year, and I have no reason to believe that this number is unusual. I have quite well-trained and effective bird dogs who now and then will find a bird, either dead or near death, that some other hunter shot but couldn't find. But these finds average less than three per year. In the research that I and my associates are conducting on the Quabbin hunters, we are asking hunters to report crippling losses so that we will have some idea of how serious an issue this is.

animals. When there is choice, not necessity, we are morally obliged to acknowledge the animals' right to "life, liberty, and the pursuit of happiness."

Why this emphasis on pain? What is it about pain that puts it at the very center of our moral universe, the criterion against which all action must be measured? For the animal rights advocates, the question is purely rhetorical—pain is so obviously bad, so universally unwelcome, that the necessity of its avoidance has all the force of a cosmic given: it is the Archimedean point from which, with the lever of moral sensibility, the earth can be moved. Lorraine Near was convinced that if vivisection were ended, science would make rapid progress. But that would only be the beginning.

> I think if we have compassion for other species, we will have compassion for one another, we will have compassion for our environment. . . . What's going to happen is that we will expand our consciousness from self to family to community to world community to other species . . . to trees and water.

Liz Simpson and her partner, Jim Ventura, were similarly committed to the idea that if we would stop being violent toward nature, not only stop hunting and killing but also stop using herbicides and pesticides in agriculture, a new era would be ushered in, an era of peace and freedom from want. For them, advocating animal rights was a way of urging a fundamental redefinition of what it means to be human.

The animal rights agenda is, in fact, less about protecting animals than it is about changing humans. By bringing humankind to a higher level of sensitivity, a more perfectly ordered and decent society will result. In turn, this will produce a healthier relationship between humans and the natural environment. Animal rights is as much about reforming human beings as it is about making animals happy or pain-free. As in other moral reform movements—temperance and the anti-abortion movements come to mind—the presumption is that certain forms of enjoyment are evil, not only in themselves, but in the sense that they cause evil to multiply.

Hunting, seen in this light, is bad because of what it does to animals and also *because it corrupts and degrades the people who hunt*. That is why hunters are such a boorish and vulgar lot. This view of hunting and hunters is virtually identical to the one held by the God-fearing folks clinging precariously to the rocky coast of Massachusetts nearly four hundred years ago: hunting gives rise to passions that might well overwhelm the admonitions to probity, circumspection, and civic virtue. Blood sport invites impetuosity and anarchy.

Animal rights is, at root, about the regulation of pleasure—the pleasure derived from the hunt, the pleasure derived from eating meat, the pleasure derived from cosmetics. Regulation of pleasure is also the exercise of power: the power to determine how people should live and what enjoyments they may seek. In attacking hunters, the animal rights advocates invoke the middle-class bias against the culture of the rural yeomanry and urban working classes (and in England, the ambivalence toward the aristocracy). The lower classes from which most American hunters are drawn are presumed to lack the virtues of impulse control, deferred gratification, sobriety, sensitivity, and tenderness. While hunting may not be the cause of these deficiencies, it is seen as an activity that perpetuates them, and in some quarters serves to make a virtue of them. Were hunting abolished, from this point of view, a giant step would be taken on the path of human improvement.

This view of hunting is curious. The case against hunting is strongest at the point of the individual animal being hunted. Its life chances are certainly not improved by the hunter's efforts. That much is clear. Beyond this, there is nothing but muddle. It is not at all clear, for example, that properly regulated hunting of the sort that has prevailed in the United States for most of this century has any negative effect on the health and viability of game populations.* Indeed, the opposite seems to be more nearly the case.

Even less clear is the connection between hunting and anti-social behavior. Are hunters more warlike? more bloodthirsty? more inclined to crime? less involved in community affairs? Using national surveys

---

*There is an argument against hunting that claims resources devoted to the promotion of wildlife have been heavily tilted toward game species to the detriment of nongame species. This tilt is evident. What is not evident at all is the allegation that this is the fault of hunters. It seems at least equally reasonable to conclude that nonhunters are thoroughly indifferent to wildlife, and are unwilling to devote even modest amounts of tax dollars to their well-being.

that have asked whether householders hunt, I have compared hunters and nonhunters on a wide array of social characteristics and attitudes. Suffice it to say here that there are no significant differences between hunters and nonhunters that cannot be accounted for by differences in social class and the cultural differences between regions of the country. Hunters are, for example, more politically conservative than nonhunters, but this is because a disproportionate number of hunters are from the South and from the mountain states, the most conservative regions of our country. In those regions, hunters are no more conservative than their nonhunting counterparts. Hunters are not, by virtue of their hunting, less civilized or more brutal or more reactionary than are nonhunters. They do feel more closely tied to traditional culture and display more nostalgia for the "good old days," but this scarcely places them beyond the pale of civilization.

But these considerations are largely irrelevant for the opponents of hunting. Class and culture divide hunters from most people who think of themselves as environmentally concerned. The alliance between hunters and the MDC faction was awkward for the agency and its nonhunting supporters and was accepted with deep reluctance by most of the MDC faction. They put the matter in perspective. Instead of allowing their prejudice against hunters to cloud the deer problem, they kept their focus on the overarching goal of ensuring the regeneration of the Quabbin forest. But the hunters were not going to be welcomed with great joy.

Though no one said so explicitly, I could not help but get the feeling that a good measure of the disquiet with hunters arose because hunters, unlike the others who manage the environment or support management efforts, are not reluctant to do so. On the contrary, they are enthusiastic and eager to "manage." Of course, for most hunters the idea of helping to manage is a small to nonexistent element in their activity. They are hunting for pleasure, for the satisfaction derived from submerging oneself in nature, becoming part of the ageless drama of predator-prey. Managers are ambivalent about tampering with nature; hunters experience little of that ambivalence because they think of themselves as one with nature, at least while hunting.

For those who think of nature as separate from humans and somehow purer than anything humans can contrive, the hunters' claim to being one with nature is akin to blasphemy. They increasingly insist that nature has a right to be left alone, that humans have no right to assert their presence in nature, since that presence invariably leads to disruption. By contrast, the hunters and the proponents of environmental management, with varying degrees of ambivalence, insist that we are ineluctably a part of nature and therefore must bear responsibility for nature's and our own well-being. For all their differences, hunters and the MDC faction spoke of nature in the language of responsibility. Their opponents spoke of nature in the language of rights.

5

Nature's Rights and Human Responsibility

For two million years we were hunters; for ten thousand years we were farmers; for the last one hundred years we have been trying to deny it all. . . .

STEPHEN BUDIANSKY, *The Covenant of the Wild*

NATURE IS CONTESTED TERRAIN. From the very beginning, organisms have jostled with one another in the search for food and safety. Even simple organisms show a remarkable inventiveness in this contest: there are the organizational feats of ant colonies; the incredible camouflage of butterflies; the stunning eyesight of the osprey. These traits, of course, aren't "invented" in the ordinary sense of the word. The osprey doesn't imagine how much better life would be if only its eyes could penetrate beneath water to spy fish for dinner and then set about to improve its vision. The completely involuntary and unbidden process of natural selection is the engine of this sort of inventiveness. We humans are part of this process too. But we are different, unique so far as can be determined, because we can invent in a fuller sense of the word. With the rest of creation, we struggle over physical space; we alone struggle over the symbolic space our inventiveness allows us to create.

The controversy over the Quabbin is a struggle over a place onto which the contending parties have projected symbolic meanings that

131

are diametrically opposed. At the root of the disagreement, though this was not what the parties explicitly debated and argued about, are two sharply divergent images of nature. Critics of the Metropolitan District Commission, for the most part, imagined nature as a realm of balance and harmony. Everything has a function and a place. If humans would only absent themselves from, or at least drastically reduce their demands on nature, nature would achieve an equilibrium among its constituent elements. Even though most of the critics stopped short of imagining lions and lambs in easy repose, they spoke of nature as a state of interdependence in which lions and lambs accepted the symbiotic need for each other. Nature is benign, self-healing, paradoxically ever changing and ever constant.

Humans are interlopers in this scene: arrogant, clumsy, ignorant intruders. For millennia, we have set about on myriad schemes to transform this or that aspect of nature to suit our fashions and transitory whims. The result has been a long series of disasters that now threaten to issue in general calamity. The threats are palpable and growing: acid rain; ozone depletion; irrevocable loss of rain forest; poisons in the air, earth, and water; global warming. The more sophisticated we become in our knowledge and techniques, the wider grows the circle of disruption and destruction. Our scientists have unleashed a monster that makes Mary Shelley's Frankenstein seem like the father in the Addams family. We need to be stopped, saved from our own foolishness and greed, so that nature might repair itself and return to some semblance of order and stability. If we will relent, if we give nature respite, "nature will take care."

The MDC faction, by contrast, saw nature in terms that are much less sanguine. In their view, nature adjusts to whatever confronts it. If the globe warms, for whatever reason, species migrate toward warmer or cooler climes as best suits them, and if the change is abrupt, the slower to respond are left behind to face possible extinction. This adjustment process is not harmonious, balanced, or in equilibrium for more than fleeting moments, geologically speaking. The process is ongoing, pushed along by continual disruptions, some from human action but others having no connection to human agency. The course of this dynamic cannot be charted with great accuracy, but to the extent that we can fathom what lies ahead, we need to ensure that the "adjustments" nature makes are compatible with our own needs. Reasonable people can disagree over what are legitimate needs, but if we want to maintain certain features of our landscape—forests of a particular type, a particular mix of warm- and cold-blooded creatures, rainfall within a certain pH range—we have to intervene regularly and systematically. We have to intervene by regulating human behavior, but we also have to intervene in the natural world in order to ensure that our needs can be met on a continuous basis. To the extent that there is equilibrium or balance, it is our creation, not nature's. Nature is random, chaotic, ceaselessly changing.

Nature is continually creating vacua as species rise and fall. Our activity also creates vacua—when land is cleared, when river courses are modified, when settlements displace native species. We may not always like how nature "chooses" to fill such vacua—locusts and ragweed are, on the whole, not nearly as welcome as deer or fern. Must we be passive about what nature throws up? Does nature know best? Best for whom? Are nature's choices any better than ours? Casting over our efforts, one might be tempted to conclude that we should refrain from all interventions except those that attempt to restore nature to its original condition, as the turkey or eagle restorations purport to do. But how do we establish "original condition"? Do we mean the time before European settlement and conquest? Do we mean the period before intense urbanization and industrialization began after the Civil War? Even if we could agree upon a moment that we would like to return to, environmentally speaking, what gives us the assurance that our image of that moment bears any serious resemblance to what was actually there? And what if that earlier, presumably more pristine state, is utterly incompatible with our present numbers or prevailing life styles? By what authority do we decree this or that arrangement "natural" and thus deserving of preservation? And even if we could settle that matter, would we not then have to manage assiduously to keep the environment that we had chosen to privilege as "natural" more or less stable?

These questions are far from new. Venerable questions tend to be accompanied by equally recurrent arguments. At least since the late nineteenth century, these sorts of questions have given rise to two

ᵥ opposed responses: stewardship versus preservation. Each position has
its own internal logic and its own prescriptions for how we should act in
and on nature. The controversy over the Quabbin is but another episode in which these two views of nature have clashed. With each
repetition of the arguments, one might hope for at least a glimmer of
some new formulation that might dissolve the differences between the
two factions. There have been glimmers, in fact. The most promising
has been Aldo Leopold's endeavor to temper our unavoidable need to
use and thus alter nature with the ethical imperatives of modesty and
circumspection. But pressures for heedless exploitation are powerful,
and even well-intentioned attempts to manage the environment too
often suffer from myopia, and not infrequently make matters worse.
Such failures virtually guarantee the continued appeal of preservationism. Leopold's very effort to find a common ground, a way of joining
the pragmatics of resource use with preservationist idealism, has been
used by partisans not so much to transcend the deadlock as to deepen
it. Perhaps we can find a way out of this impasse by first attending
closely to the language the two sides use to advance their respective
positions.

The Language of Rights

The opponents of the MDC were essentially preservationists. Some
would have liked to see as much of the earth as possible returned to
something approaching its condition before the industrial revolution
began to transform the global landscape. Others were grudgingly reconciled to the larger losses and wished simply to preserve as much as
possible of what is still more or less "natural" and to restore tracts like the
Quabbin to as near their original condition as can be achieved with
present resources. As we saw briefly in chapter 1, the preservationist
position has been around for at least 150 years, from the time of Thoreau.
But an element in this argument that Thoreau only hinted at has now
become the centerpiece of the contemporary preservationist position.
Preservationists increasingly couch their position in terms of the language of rights: specifically, they would extend to nature the right to be
left alone. This emphasis on rights has given the preservationist move-

ment new energy and a whole new vocabulary for countering the allega-
tions of unrealism and impracticality that have continually plagued their
cause.

The fullest and most explicit use of the idea of rights comes from the
advocates of animal rights. Even though many of the MDC's critics held
back from a full embrace of the animal rights position, they nonetheless
found the language of rights compelling. Long accustomed as we are to
basing solicitude for others on the presumption that they have rights
we are bound to respect, it would seem logical to extend rights to other
creatures, even the whole of creation, for whom we feel similar solicitude. Whether this sympathy proceeds from an identification with the
capacity to suffer and feel pain, as with the animal rights advocates, or
from the expansive notion that all life is equally important because each
life form is inextricably linked to all others, the point is the same: caring
requires the grant of autonomy, the right to be left alone.

Thinking of animals and nature in terms of rights has also no doubt
received a boost in recent years from the broad tendency in our culture
to litigate. It seems likely that the shift toward the courts as the bat-
tleground for public policy is but one result of the cynicism and distaste
Americans have for politics and politicians in general.* Add to this a
key provision of the Endangered Species Act of 1973 that specifically
enables interested citizens to sue to stop a development or other action
that they can demonstrate might harm or jeopardize one or more
species on the federally compiled endangered list. Suddenly, otherwise
mute animals and plants can acquire self-appointed spokespersons
who, in the name of the threatened species, can ask the courts to stop a
dam, a subdivision, a logging operation, or a deer hunt. Though the
courts impose their own sometimes arcane standards of proof, and the
burdens of such suits can sometimes be quite large, the courtroom has

---

*Students of the animal rights movement have found activists extremely disillusioned with con-
ventional political processes, not least because ordinary politics requires give and take and compromise that, to the committed animal rights advocate, cannot help but be seen as "selling out."
By contrast to the inevitable ambiguity and fudge of legislative or electoral wrangling, the language of rights has compelling simplicity and clarity. For two careful recent studies of animal
rights activists, see Susan Sperling, *Animal Liberators*, Berkeley, CA: University of California
Press (1988); and James M. Jasper and Dorothy Nelkin, *The Animal Rights Crusade: The Growth of
a Moral Protest*, New York: The Free Press (1992).

nonetheless offered environmentalists a more nearly equal footing in their battles with developers and exploiters than have the halls of Congress.

Resonant with deep veins of our culture and promising positive results, the language of rights is increasingly used in environmental policy deliberations. But before we all leap on the bandwagon, it would be wise to reflect on some of the less obvious implications of this extension of rights to animals and to nature itself. First, let us briefly consider how the language of rights came to be applied in human affairs.

Autonomy based on the right to be left alone arose in Western thought as a corollary to the idea of individualism, the conception of the individual as the singular vessel of moral reasoning and thus the bearer of responsibility for his or her own acts. Autonomy could be granted to individuals, it was reasoned, because individuals could frame a contract, a "social contract," wherein they would agree, for the good of all, to honor each person's rights. Should the contract be broken, should an individual's rights be violated, the state would seek to rectify matters, either by restitution or criminal penalty. Rights, then, are a human invention, not heaven-sent. They are constituted by a political authority whose purpose it is to enforce the laws establishing those rights and to punish those who transgress. Within the framework of these laws, individuals are presumed to be free to pursue their own goals and life plans. The laws ensure, at least abstractly, that order is maintained and that conflicting claims can be settled in routine and peaceable fashion.

Human affairs, of course, are rarely so simple. This bland account masks a very turbulent four-hundred-year history in which the implications of these general principles have been thrashed out. Indeed, the thrashing continues to our own day. As a society, we have barely settled the argument over whether blacks are due the same rights as whites, and we are a long way from accepting the full implications of what such rights entail. We are also still grappling with the proposition that women are entitled to the same rights as men. We are not even close to accord on the proposition that fetuses are entitled to the same right to life as persons. Nor are we settled on whether the mentally ill or the comatose or the handicapped have the same rights as the rest of us.

At issue in such debates over the extension of the claim to rights is the deeper debate about who should be considered members of the community covered by the blanket of rights. The historic arguments over whether nonproperty holders or slaves had rights were, like the contemporary debates over fetuses and animals, arguments about who is to be included in the social contract.

Membership is historically variable, but the general trend is toward expanding inclusiveness. Momentum, as they say in sports broadcasting, would seem to favor extension to nonhumans. This is the burden of Roderick Nash's history of environmental thinking, *The Rights of Nature*. But we should proceed cautiously. The most troubling and problematic extensions of rights come from the attempts to include as members those whom we cannot presume to have the capacity to decide autonomously upon life plans. Fetuses, for example. What does it mean to grant rights that protect the autonomy of an organism that clearly is not autonomous?

Deer plainly have a clearer claim to autonomy than do fetuses, and yet they share one important quality with the unborn: they cannot speak on their own behalf. This means that those who would presume to represent their interests have to establish a claim to knowing what those interests are. To do so is no simple matter. In an era when persons with congenital defects sue their parents for "wrongful birth," the assumption that each creature has a desire to live and to avoid pain contains demonstrable contradictions. What about a life overbrimming with pain? Are we prepared to say that all deer have a right to be left alone, if we know that by doing so we will condemn large numbers of these deer to lives as malnourished, stunted, and deformed creatures certain to die torturously slow deaths?

Though autonomy may seem intuitively compelling, to use it as the basis for extending a legal protection to classes of organisms like animals or trees requires that we have some reasonable way of establishing who is entitled to speak for them. And if these organisms need spokespersons whom they cannot themselves choose (and thus cannot possibly dismiss in favor of some other set of spokespersons pressing for a different set of interests), does this not seriously undermine the prior claim to the right to be left alone, the right of autonomy? Why should we construe the animal kingdom as bent on the avoidance of pain, an interest we must respect? Is there not ample evidence of nature producing, on its own, a surplus of pain? How then can the pain we humans impose, as in the case of hunting deer, be viewed as somehow excessive or "unnatural"? There is no compelling proof that the hunters' construction of what is best for deer is any less coherent or less firmly grounded in a body of information and evidence about deer and their environmental needs than the construction of the deer's interests and needs given by the animal rights advocates.

The claim to rights involves more than the presumption of autonomy. A capacity for competent choice and reciprocity is also required. To make the extension of rights coherent, we must assume that the bearer of rights can meaningfully submit to the laws protecting its and everyone else's rights. For this reason juries must find defendants sane before they can find them criminally liable. If people are not in charge of their faculties, their responsibility is diminished and their claims to autonomy reduced. In the matter of animals, what does it mean to extend rights if we cannot reasonably expect the animal to reciprocate, to heed the claim to rights of other members of the biotic community? More specifically, how can we concede to deer the right to an unfettered life if we cannot presume that the deer will conduct themselves in ways that honor other creatures' rights to the same unfettered life? Can we meaningfully say, as we could about humans, that deer have broken the social contract when they overbrowse their habitat, thereby making it unfit for a wide range of other plants and animals? Aren't they merely doing what deer always do? Certainly they do not possess malevolent intent. They are just being themselves: creatures that will go on eating and reproducing until they run out of food. Does the fact that the deer have no choice but to behave as they do mean that we cannot intervene to head off the larger ill effects of overbrowsing? Why are the deer's rights more compelling than the rights of the oak saplings?

These questions create serious problems for those who wish to use the language of rights to protect individual animals or as a basis for a more encompassing environmental ethos. In the face of such questions, the language of rights can remain a plausible way to think about the interaction between humans and the natural world only if there exists in nature an analogue to the social contact in human affairs. There has to be some mechanism that balances out the claims of one animal or one species with all other claims. Otherwise, the language of rights is meaningless. Without an ordering principle akin to the rule of law, the only rights that could possibly come to prevail would be those claimed by the mighty: might makes right.

This is why the advocates of rights for nature are drawn to images of nature that stress balance, harmony, and equilibrium. In the absence of some clear and powerful regulatory force, nature would be Hobbesian— a war of all against all—and it would make no sense to talk about the rights of animals or trees. But if everything in nature were granted the right to be left alone and, on fulfillment of this condition, things became balanced in such a way that an equilibrium was achieved, then one might reasonably conclude that each element was, as it were, behaving in a law-abiding way and that each element's rights were being respected. If nature's true character is one of balance, then the guarantee of rights amounts to the protection and preservation of a harmonious nature.

In effect, preservationists face the same dilemma that confronts the advocates of an unregulated economy. If humans are assumed to be propelled by self-interest, then it can also be assumed that animals and trees are self-interested, that is, they are single-minded in their determination to stay alive and to reproduce themselves. Somehow, there has to be a way to balance out everyone's pursuit of self-interest such that a version of the common good can emerge. Classical economists, following Adam Smith, thought they had found just such a transformative agent in the free market. The free market is governed, so it is argued, by the famous "hidden hand" of supply and demand, by which everyone's selfishness is transmuted through countless transactions into the greatest good for the greatest number. Those who assert that nature has rights are similarly in need of some hidden hand that transforms maximizing individual organisms into an orderly community in which rights are protected and there is something for everyone.

The MDC critics relied essentially on two versions of the hidden hand. One is rooted in the venerable tradition of field biology and natural history that proceeds from the assumption that the natural condition of things is dynamic equilibrium. The other is rooted in an equally venerable tradition of spirituality. Let us examine each in turn.

Field biology formed the initial framework for modern ecology and has informed both scientific and popular thinking about nature and the environment for well over a hundred years. Biologists and ecologists still use the model of dynamic equilibrium extensively.* In the nineteenth century, many naturalists drew an explicit parallel to the field of economics. Books and articles regularly used the phrase "nature's economy" or "natural economy." Nature was portrayed as a vast self-adjusting system of supply and demand—if the reproduction of rabbits rose in one year, the population of foxes would respond in the next. Nature was efficient—even extravagant plumage or stunning coloration was exactly what was required for that creature to fit into its place in the scheme of things. Just as social Darwinists found justification for the dominance of white Anglo-Saxons in Darwin's theory of natural selection, so too did commentators find comfort in the belief that they had discovered in nature economic principles by which human society could be ordered as efficiently as nature orders its affairs.

The persistent notion that nature is self-regulating is held with much the same conviction that led nineteenth- and early twentieth-century economists to insist that, under laissez-faire conditions, depressions are an impossibility. So, for example, U.S. Park Service scientists steadfastly maintained that the elk herd at Yellowstone National Park could not possibly grow larger than the habitat would permit. Because of this belief, as Alston Chase makes painfully clear in his book, *Playing God in Yellowstone*, the park service stood by while the elk proceeded to devastate the park, rendering large areas of it unfit for

---

*In its most ambitious and sweeping form, the model appears as the Gaia hypothesis, as its proponent, British scientist and eclectic James Lovelock, dubbed the idea. Lovelock argues that the earth is one vast system in equilibrium. A change in one aspect is compensated for by a change in another such that the whole thing continues. Particular life forms ebb and flow, new life forms are continually being thrust forward, some of which thrive, others of which perish, often without a trace. The whole is a giant living organism in which all parts and pieces have a place, though none is indispensable, including humans. Life goes on, with or without dinosaurs, elephants, rain forests, or Homo sapiens. See his *Gaia: A New Look at Life on Earth*, New York: Oxford University Press (1979).

the other species that had depended upon the willows, aspen, and other growth that the elk monopolized. Despite mounting evidence to the contrary, many biologists and naturalists continue to hold that there are, in V. C. Wynne-Edwards's words, "self-regulating systems in populations of animals."[1]

As we have seen, this sense of nature as a balancing act that efficiently produces optimal mixes of species was broadly shared by the critics of the MDC, especially those who emphasized the importance of the ecosystem. Devoted amateurs like Ray Asselin and Bill Granby regularly invoked images of balance and equilibrium when describing the wonders of old-growth forests. Nature, in their view, can produce pure water much more "cost-effectively" than any system humans contrive. If we would keep our tamperings as close to zero as possible, we would spare nature many grievous injuries and we would also save our own precious resources: time and money. And since the natural order is self-regulating, we needn't worry about intervening on behalf of this or that plant or animal. Nature is not only efficient, it has its own elemental evenhandedness: nature plays no favorites and thus can safely be left to its own rhythms.

While most of the critics were willing to go along with this way of seeing nature, it was clear from their remarks that they had not based their convictions on the empirical evidence derived from the work of biologists or naturalists. Their sense of things derived more from a diffuse spirituality than from a carefully worked out system of ecology. Interestingly, none of the critics I interviewed explicitly invoked God, but many were plainly caught up in a view of nature that was commingled with ideas of the supernatural. Many of the MDC critics were drawn toward an eclectic mix of Native American myths, Eastern religions, and more conventional Western deism. Present, too, were elements of what has come to be called eco-feminism.* Taken together, this mixture seems to fit comfortably into a family of beliefs that the historian Catherine Albanese analyzes in her recent book *Nature Religion in America* (1990).

Like their predecessors in the nineteenth century who found in

---

*For a sympathetic but critical discussion of the range of views that found expression among the MDC's critics, see Carolyn Merchant, *Radical Ecology*, New York: Routledge (1992).

nature's exquisite work models of innocence and harmony that humans would do well to emulate, many of the MDC's critics were worshippers of nature. A transcendent force was at work in the cathedrals of the forest, a force that regulates, orders, and balances. No one embodied this belief more fully than Peter Gomes. His public lectures could easily serve as an archetype of "nature religion." For him, nature simply is harmonious. We just need to let it be and learn from it all we can. As Peter said, with intense urgency, we can reclaim our place in the Garden if only we stop meddling. For Peter and several others who shared his convictions, the Quabbin could be a beachhead, a place to start the long process of reforming ourselves and allowing nature to heal itself. Others of the critics were less visionary, though no less spiritual, and were willing to concede the grim reality of sweeping environmental degradation. They hoped, more modestly, that areas like the Quabbin could be left alone to become places where people could retain vibrant contact with untamed nature.

But what if equilibrium and harmony are not the original condition of nature? What if disruption and tumult are the real nature of things? Then, clearly, the granting of rights becomes problematic. If species are opportunistic, if they rush in to colonize in the wake of some disturbance or other, heedless of existing tenants, the right to be left alone evaporates. The presumption that honoring such a right will make for a healthier and more stable environment also evaporates.

Twenty-five years ago, the biologist Garrett Hardin wrote a short essay, "The Tragedy of the Commons" (1968), that has become a minor classic of environmentalism.[2] In this essay, Hardin laments our individualism. As individuals, he shows, we pursue self-interest, largely heedless of the cumulative effects of our individual actions. In the process, the commons, here a metaphor for the environment, gets systematically neglected. Historically, the real commons of English and New England villages were subjected to overgrazing and neglect. Though the maintenance of the common was in everyone's general interest, it was in no single individual's personal interest to moderate his or her use of the common. The result was tragedy—depletion that left everyone the poorer.

The critics of the MDC were quite prepared to accept this indictment of human shortsightedness and greed. Indeed, they saw the events at the Quabbin as a case in point: humans pursuing short-term advantage at the expense of what could be a marvelously rich and beautiful natural area. But they resisted the implication that the deer, for example, might be at least as greedy and shortsighted as we humans. The critics insisted that if the deer overran and overbrowsed, it was because of things that humans had or had not done, not because the deer were out of control. In a similar vein, Desmond Morris, the author and animal rights advocate, argues in his recent book *The Animal Contract* (1990) against the notion that wild animals are vicious. He writes, speaking of what we can learn from those who make pets of exotic wild animals such as leopards:

> All too often in the past the savagery of such animals has been caused by the brutal way they have been treated by humans rather than by some inherent viciousness in their characters. The 'savage beast' is largely the invention of the cowardly big-game hunter. (65)

When I proposed applying Hardin's analysis to the behavior of the deer at the Quabbin to Lorraine Near, one of the leaders of the animal rights opposition, her response was wistful:

> I have a pretty spiritual attitude about that [animals' "selfishness"]. I think we should always be evolving towards a more peaceful state with our nature, with animals, with ourselves.

I do not think it was careless speech when she said "we *should* be evolving" instead of "we are evolving." To allay doubts, it is necessary to will that nature be headed in the right direction, in the direction of "a more peaceful state." We humans may be selfish and shortsighted, and thus prone to destroying the environment, but this cannot be said of animals. They are innocent.

The MDC critics portrayed humans as aliens, almost as a conquering horde. The best we can do is leave things alone. If there is managing to do, it is almost entirely self-management that is necessary—controlling our impulses to dominate and manipulate; reducing our population growth; restraining our search for gratification; learning to accommodate to nature's cycles. Because we have ruptured the organic link between ourselves and nature by our long history of manipulation and exploitation of nature (in Desmond Morris's phrase, we have violated the "animal contract"), we must radically alter our course. We must accept the fact that we have no right to continue to manipulate and exploit. Nature and humans alike must be protected by a bundle of rights, leaving to nature, or the heavens, the task of adjusting claims and balancing things out. In this balancing act, each living thing has equal standing, is equally precious.

This emphasis on the individual organism, especially prominent among animal rights advocates, had created some tension, as we've noted already, within the ranks of the MDC's critics. For those whose commitment was to the ecosystem as a whole, the radical individualism of the animal rights position was unattractive. But the two groups were able to make common cause, despite this tension, because both shared the sense that nature, in the end, produces harmony and balance. Thus, for the animal rights advocates, animals must be free to live out their life plans *so that harmony can be restored to nature.* For the more traditional environmentalists, it was the dynamic balance itself, not the unimpeded life plans of individual animals, that was central. Despite this quite important difference, both saw human intervention as the principal source of disruption and thus wished to have nature protected by a blanket of rights that would oblige humans to treat it more reverentially.

The Language of Responsibility

Those who supported the hunt held a very different view of nature. While certainly agreeing with their opponents that humans have built a sorry record of environmental neglect and abuse, they resisted the impulse to lay blame for everything at the human doorstep. Nature is, in this view, anything but benign. If we cherish some feature of our landscape, we not only have to protect it from thoughtless human activity, we also have to protect it from natural forces that are, literally, indifferent as to the value of this or that. We could, of course, take whatever nature casts up.

This, presumably, is what Homo erectus did for much of prehistory.

It is what the rest of creation still does. But given who we are, given what nature has endowed us with, can we really say that it would be natural for us to sit back and passively accept whatever comes our way? Aren't we obliged, by our very nature, to seek advantage by utilizing our capacity to reason, to modify, to invent? Can we escape this imperative any more than deer can escape their drive to breed, even when it brings them to the point of exhausting their food supply? If nature presents us with dry seasons, aren't we entitled to use our imaginations to contrive ways of storing water when it is plentiful for use when it is scarce? Must we accept the higher mortality and shorter life spans that would surely return were we to refrain from storing water (or food or fuel)? Don't we have a responsibility to one another to intervene?

Obviously, the MDC faction thought so. Thinking so produces a corollary: we also have a responsibility to nature. Once we begin to intervene, we cease to have the moral luxury of refusing responsibility. We have, for better or for worse, altered the face of nature—there is simply no place on the globe that remains pristine. Our past intrusions, intentional and unintentional, have bequeathed to us, the living, the imperative of continuing intervention. To now say "hands off" is to leave all that we have nurtured and ignored, all that we have loved and all that we have abused, adrift in a force field that knows no mercy and has no rhyme nor reason. To speak of the right to be left alone in this context struck the MDC faction as utterly irresponsible. Most charitably, it was whimsical foolishness.

The MDC faction did not believe that, left to its own devices, nature would provide. Sometimes it does and sometimes it doesn't. They also did not believe that nature's intrinsic character can be described in terms of balance, dynamic or otherwise. Perhaps, in the longest of views, there might be some sort of vast self-adjusting principle of the sort that Lovelock proposes with his Gaia hypothesis. But the notion that this adjustment produces an environment that we would find acceptable, much less bearable, is far from self-evident. Without active and unremitting intervention in and manipulation of nature, we would almost surely become an endangered species. Without circumscribing this intervention with the exercise of responsibility, of stewardship, we are equally sure to become endangered, along with most everything else. However burdensome it may be, we have no option but to manage, to assert that particular arrangements and mixes of flora and fauna are preferable to other mixes and arrangements. The alternative is chaos and virtually certain disaster.

The noted ecologist Daniel Botkin has contributed significantly to this view of nature. His book *Discordant Harmonies* (1990) was making the rounds among MDC personnel just as I was conducting my interviews. Botkin, though far from being an apologist for management, discusses a number of instances that reveal how nature does not produce balance, at least not automatically or inevitably. Whether this is the result of the cumulative distortions human activity has imposed or the result of the essential nature of nature itself remains an open question. In either case, the implication is the same: we have no choice but to intervene to preserve those features of our environment we value, whether for their intrinsic virtues or for their utility. Botkin concludes:

> Nature in the largest sense is a system that has varied over time and space at many scales. We are left with the realization that we have the power to change the biosphere, and we are forced to make choices; nature in the large does not provide a single simple goal. There are many themes in nature's symphony, each with its own pace and rhythm. We are forced to choose among these, which we have barely begun to hear and understand. (183)

Nature, from this vantage point, doesn't manage. Like Old Man River, nature "just keeps rollin' along." Individual species jockey for position. Some win, some lose. Over time, winners and losers frequently trade places, though not in any way that could be described as blind justice, much less balance. At the Quabbin, this process has boiled down to a choice between humans managing or deer managing. Either way, the forest will be given a particular character that in no serious or important sense will ever resemble that of some imagined original forest. Even if we did turn our backs on the Quabbin to let nature take its course, the results would reflect the overwhelming artificiality of the reservoir, the unusually high deer population, and the effects of acid rain, ozone depletion, gypsy moth infestations, and countless other influences that have accumulated in the ten thousand years or so of human presence in the area. Why prefer whatever this outcome would be to the outcome we can produce by deliberately intervening to forge a balance between deer and vegetation? Why is our design less worthy than the random outcome of impersonal natural forces?

To emphasize responsibility is to resist viewing nature as composed of discrete, atomized individuals or discrete separable tracts. Just as John Donne could write in response to the growing individualism of his day, "No man is an island," so the M DC faction rejected the idea that the Quabbin, as large as it is, could be treated as isolable and separate, a world unto itself. It can't be made an island of wilderness. To the extent that it seems wild, it is wild because humans have actively made it so, in accord with what, in our culture, we have come to think of as wild. And just as surely, our continued action will be required if it is to remain wild.

There is considerable irony in this view. If we have to manage and manipulate to keep something appearing wild, what does wild mean? Is it wild at all? More than just irony is involved in such questions. If we accept the argument for responsibility, we have to confront some very troubling questions. Who decides what is wild? Experts? Politicians? The public? Apart from the question of who has the power to decide what is wild (or what is sound environmental policy), there are the ambiguities that arise when the inevitable disagreement occurs over policy and its goals. Should we manage for sustained yield of resources? Should we strive to reproduce a precontact forest? Should we work to maximize biodiversity? Should we endeavor to avert extinction of a species, even at considerable cost and inconvenience to us and possible disadvantage to other species? Such questions reveal the ambiguities of using power. Let's consider them briefly.

Who among us should we trust with the power to decide how nature should be managed? The critics of environmental management, as we have seen, are quick to challenge the credentials of any and all candidates. Professional foresters have, they allege, a bias favoring commercial standards: tall, straight trees from which marketable wood products can be readily obtained. Wildlife biologists, too, are seen as captives of a harvest mentality, this particular variant supported by those pressing for the consumptive uses of fish and wildlife. Similarly, politicians are easy prey for special interest groups, the most powerful of which are almost always those that stand to benefit most from short-term gains. In brief, there is no genuine disinterestedness, no objective or neutral authority whose exercise of power wins universal assent.

The MDC faction was quite aware of this problem. Most of them had fought many battles in behalf of what they believed was sound environmental policy, battles that had placed them in direct opposition to the people officially in charge of setting policy or, what often amounts to the same, of allocating resources to implement a policy. They found it ironic, but not amusing, to suddenly be "the enemy." From their point of view, MDC policy was the least heavy-handed policy possible, given the mandate to keep good clean water flowing to consumers in Boston. The MDC foresters were trying to produce what was generally accepted as a "typical New England forest." Ruling out nonnative species and exotics, they aimed to produce as varied a mix of species and age classes as could reasonably be sustained. This, to their way of thinking, was compatible both with sound environmentalism and with sound resource management.

However defensible their goal, the fact remained that the MDC was in charge and made its policy with little or no broad public involvement. As we have seen, there were a number of citizen advisory groups with whom the MDC met regularly and informed of general policy directions. In this sense, the public had access and input. But for all intents and purposes, the real public was powerless. As countless studies have shown, these kinds of arrangements almost invariably play into the hands of "special interests." With western rangelands, it is cattlemen whose interests prevail. It is no different with the national forests—timber companies prevail. The celebrated instances, as in the case of protecting the old-growth habitat of the northern spotted owl, when grass-roots environmentalists win out over commercial interests are so rare as to be exceptions that test the rule.*

---

*Not only are victories rare, even when they occur they are all too regularly ambiguous. Portions of habitat are saved but turn out to be too small or too fragmented to provide the protection sought. Or the protection comes too late and all that is achieved is a slowing of the decline. There have been, of course, some spectacular recoveries, some by direct human agency, some by virtue of the resilience and adaptability of a plant or animal species. But on the whole, the proponents of management have to deal with a tally sheet that would get most other sorts of managers fired.

Considering this track record, what was truly amazing about the Quabbin was that, with the exception of sport fishermen, special interests had made no inroads. The MDC had been able to withstand the pressures from loggers as well as the concerted efforts that had been made from time to time by proponents of recreational boating and cross-country skiing, to name only two interest groups. The logging that went on at the Quabbin under the close supervision of the MDC foresters was carried out with restrictions that, were they applied generally, would be hailed by environmentalists as an almost undreamed of standard of environmental protection. But for those who wanted preservation, not protection, who wanted no use rather than restricted and conservative use, this was not good enough. Lacking effective power within the policy-forming process, they could not help but see loggers and fishermen—and now hunters—as having subverted and corrupted the MDC.

While the Quabbin is, I daresay, the weakest possible example of special interests suborning resource managers, and the MDC critics' case a poor one, it is nonetheless true that those who defend the stewardship position have much to answer for. The preservationist argument may be full of contradictions and impractical in the extreme, but these weaknesses do not diminish the force of the charge that stewardship has all too regularly been a cover for environmental exploitation. However compelling in the abstract, the stewardship position has been more rhetoric than substance. If stewardship is to be an effective alternative to preservationism, the stewards are going to have to find the backbone to withstand those who are intent on short-term maximization and immediate pleasures. Obviously this is no easy matter. There are powerful industries in our society that are devoted to resource exploitation and they are backed by consumers—us—who want cheap water, inexpensive food, abundant building materials, and space over which to spread themselves in suburban enclaves.

Managers like those responsible for the Quabbin are far more comfortable in the woods than in hearing rooms and legislative assemblies. They would prefer the close observation of trees or tree frogs to close attention to the shifting alliances of politicians and the ploys of special interests. But with no clear, uncontested science to appeal to, and with their seeming allies badly split on the very meaning of nature and our relationship to the natural world, stewards have no choice but to become political. When they do, they can expect sharp challenges to their authority and their motives. It is not surprising that they can quickly become disheartened and disillusioned. Power can not only corrupt, it can demoralize.

The language of responsibility is, by the very nature of things, conditional. By contrast, the language of rights is framed as universal. Responsibility means deciding when to act and when not to act. With rights, there is but one decision, the decision to grant the right: thereafter, the matter is settled. Context, ambiguity, uncertainty—all are reduced if not completely set aside. If nature takes care and we agree to respect its right to be left alone, we are absolved of responsibility for the outcome, just as the entrepreneur can shrug off the accusation of fostering poverty by claiming that the market, not he, sets wage rates and determines the demand for labor. But those who argue for stewardship, for the exercise of responsibility, not only have to contend with powerful and almost always competing interests, they have to be answerable for the consequences of their choices.

It would be one thing if the consequences could confidently be predicted. We all know, though, that this is not possible. Again, the critics of management make a telling point: we just do not know enough to predict the outcomes of our interventions. Even simple modifications, innocuous in themselves, can set off complex chain reactions that produce unwelcome results. Couple our still-slender store of knowledge with the fact that we have virtually no control over many of the factors that we do know about, and you have decision making under conditions of high uncertainty and maximum vulnerability. A choice that seems plausible under one set of conditions can become a nightmare with even the slightest shift in those conditions. Sometimes the shift comes from swings of public mood, sometimes from new information or new theories that cast everything in a fresh light, sometimes from new hazards (like acid rain); all that is certain is that shifts will come and new choices will have to be made.

For now, the MDC thinks that it is wisest to create as diverse a forest as possible. Homogeneity, whether of age classes or species, is both biologically risky and aesthetically boring. The MDC foresters are reasonably confident that they can produce a level of biodiversity that will be superior to that which would be likely to occur were they to leave the forest completely to its own devices. Whether or not this claim is warranted, their approach seems prudent. Risks are blunted by being spread out over a number of species and a range of age classes of those species. If a sudden climate change occurs, for example, there will almost certainly be some trees that can adapt even if others perish, thus keeping the watershed forested.

As reasonable as this aproach seems, though, it raises persistent doubts. Committing to one type of forest means being prepared to tamper with many things to keep the whole system on a more-or-less even keel. It means fire suppression measures. It means controlling deer populations. It may mean controlling beaver. And on and on.

To the extent that we prefer or need balance, from this point of view, we ourselves must establish the parameters and regulate the elements that we think ought to be in balance. And it is then we, and we alone, who must accept the responsibility for the outcomes, whatever they might be. This is not a comforting prospect. It is easy, in this light, to understand how earnest, thoughtful, and intelligent people can be drawn to the conclusion that the best course is to "let nature take care."

Rights and Responsibilities Reconsidered

It may be possible some day to establish beyond doubt, in the way that we know that the earth is not flat, the true character of nature. But for now, and very likely for a long time to come, we will have to live with profound uncertainty. Our models of how nature functions are still rudimentary and subject to intense dispute. As we have seen, ecologists no longer think in terms of biological communities as having a "normal" or fixed character, departures from which can be measured and returns to which can be charted. Without such standards against which to judge the condition of ecosystems, it is hard to insist that one condition is more "natural" than another. As the ecologist David Ehrenfeld puts it in his book *Beginning Again: People & Nature in the New Millennium* (1993), "If the organismic theory of communities were still dominant, if the idea that communities have a normative, equilibrium position, a balance point, were still widely accepted, then the idea of ecological health would pose few problems" (140).

The absence of generally agreed upon definitions of "normal" means that we will not be able to resolve disputes such as the one that engulfed the managers of the Quabbin by appeal to some widely shared and objectively arrived at scientific standard. Though more research ought to be done, more research by itself is no more likely to avert conflict than it is to usher in a new era in which we can at last live in harmony with the natural world.

At least as regards the fate of the Quabbin, the plain fact is that the critics may well be correct: the water supply may not be compromised by a hands-off policy and management may be unnecessary and even harmful. No matter how the deer sculpt the forest and tilt the eco-system of the Quabbin, nature may be resilient enough to compensate and adapt. Eventually, deer and their food supply will come to some sort of standoff with many fewer deer and oak than are now present. But water will still flow, and it may be as well cleansed by whatever ground cover happens to be there as by the current forest cover. The only human intervention necessary would be to ensure that people observe strict regulations on use—regulations of the sort the MDC had in place from the start.

As we have seen, several within the MDC faction admitted to genuine curiosity about what would happen if the critics' view prevailed. What would the reservation come to look like? Would it be the lush, varied ecosystem that the critics imagined, a scene suggestive of what southern New England was like before the Europeans came? Or would it be hideous—monotonous stretches of thinning, aged trees surrounded by ferns, home to a steadily narrowing array of flora and fauna? Either outcome is plausible.

With such diametrically opposite results equally likely, and with "balance" by no means synonymous with benign, it would seem that the language of rights has little finally to offer. Since outcomes are uncertain, a hands-off approach will be politically unacceptable except in those rare instances in which we are indifferent to the result. So long as we care about what happens, whether because a vital resource like water is at issue or because an aesthetic ideal is involved, majorities are not likely to rest content with granting rights to nature.

If we can take little comfort in the idea of rights, responsibilities, too, have their limits. Most important of these is the problem of accountability. The advocates of rights for nature avoid the problem for the simple reason that nature is accountable to no one but itself. Though humans do not have quite so free a hand, our accountability is more apparent than real. In matters environmental, the true judges, the real winners and losers, are never present when they are most urgently needed: they are yet to be born. As a result, our sense of accountability is, at best, weak. Especially in cultures like our own, cultures that emphasize the present and deflect anxieties about the future by a deep faith in the certainty of progress, the links to the future that constrain action in the present range from tenuous to nonexistent. We wittingly and otherwise have arrogated to ourselves the right to pass on to our progeny the myriad costs of our own indulgences.

To some degree, this has always been so. The earth one generation bequeaths to the next is not the same earth it inherited. But in the last two hundred years or so, the difference between inheritance and bequest has grown enormous. Though some aspects of this difference are almost certainly beneficial, many, obviously, are not. With the introduction of long-lived toxins, among which are our radioactive wastes, we may well have already condemned our progeny to lives more fraught with danger than any yet faced by human beings. But since the unborn are voiceless in the matter, we feel scant accountability.

Without accountability, talk of responsibility is cheap, just as talk of rights in the absence of enforceable law is empty. This is what Aldo Leopold was getting at when he urged that we "think like a mountain." Environmental accountability, if it is to mean anything, has to be anchored in a view that extends well beyond our lifetime. It may be asking too much for us to think in geological terms, like a mountain, but it is not too much to expect that we judge our actions by their implications, as best we can make them out, for the next several generations. At the very least, we should do as little as possible that is not, insofar as we know, reversible.

Mistakes will be made. Of this we can be certain. If, however, our mistakes can be undone or at least ameliorated, we can be said to have been accountable, to have behaved in a stewardly fashion. If our mistakes are irrevocable, by contrast, we clearly have been irresponsible. We have tied the hands of future generations, squandered their possibilities.

Had the controversy over the deer at the Quabbin focused more on what the MDC might do to ensure the widest range of choices and the greatest flexibility to respond to new information and new theories, the debate might have produced the sort of constituency that the ethos Leopold was articulating so desperately needs. After all, the deer hunt as well as MDC forestry meet the test of revocability—nothing the agency has done is irreversible. Even the reservoir could be drained and the dam demolished. In a hundred years or so, none but very close observers would know what had transpired in the Swift River Valley. Sadly, the divisions within the environmental movement revealed in the controversy may not be undone as easily. Competing claims about science got combined with myth, prejudice, and defensiveness, and the resulting mixture exploded, badly dividing people who ought to have been allies in behalf of a more enduring and enlightened environmentalism. Nature may not be benign, as some of the critics imagined it to be, but it might well be more forgiving than the humans who fought over how nature ought to be understood.

6

Constructing Nature

But nature is a stranger yet;
The ones that cite her most
Have never passed her haunted house,
Nor simplified her ghost.

EMILY DICKINSON

THE DULL STACCATO of shotgun fire was intense in the first hour or so of legal shooting. After this initial flurry, it declined dramatically, interrupting the blustery cold day at odd and widely separated intervals. No one knew clearly what was going on within the reservation. The hunters had a good close-up view of things but could only judge what was happening by the number of shots they heard and by the brief encounters they might have with fellow hunters. After all the talk that had preceded the hunt, and the speculation about the high number of deer and the proverbial big bucks in the area, the hunters might well have felt anxious lest they not be successful. The volume of gunfire early on could only have added to their level of expectation. The stakes were high and the rewards comparatively low. Getting a deer on that first morning would be nothing to brag about, given the general consensus that deer were not only abundant but also quite tame. But not getting a deer was to invite almost certain ridicule from one's buddies. Were the gunshots killing deer? No one could be sure.

Outside the reservation, MDC officials huddled, as much to share their anxiety as to turn back the raw cold of the morning. They had much to be anxious about. They had invested a lot of themselves and their professional reputations in this hunt. One hunter fatality or dead eagle and there would be hell to pay. With the psychological stakes for the hunters heightened by all the hype, the risk of accidents was not small. And what if few deer were taken? What would that say about the MDC's estimate of the deer population?

Two-way radios crackled, connecting MDC personnel at the various check points to one another, but little hard evidence was forthcoming in the early hours of the hunt. Adding to the air of anxiety was the glare of publicity. TV news crews in brightly painted vans were cruising up and down the Daniel Shays Highway, looking for a story. Deer hunting is scarcely news in and of itself, even deer hunting on land where it had formerly been prohibited. The potential news that brought out the reporters and their cameras was the possibility that animal rights activists would try to interfere with the hunt or to confront hunters.

At several of the public meetings preceding the decision to go ahead with the hunt, demonstrators, many of them from Citizens to End Animal Suffering and Exploitation, had set up informational picket lines to make plain their dismay at the prospect of killing deer. Similar hunts in other parts of the country had been greeted by militant protesters. Moreover, just days before opening day, CEASE had tried unsuccessfully to block the hunt in the courts. MDC officials were worried that, thus frustrated, animal rights activists might try to interfere with the hunt or stage confrontations with hunters that could easily turn nasty.

Indeed, over the weekend before opening day, the tires of several vehicles parked at access points to the reservation had been slashed. Though no one claimed responsibility for the vandalism, authorities speculated that the vehicles belonged to hunters who were doing some last-minute scouting and that the tire slashers were people determined to stymie the hunt. In response, the MDC had decided to discourage protest with a massive police presence. The numbers of police vehicles added to the tension and excitement.

As it turned out, only a couple of dozen protesters showed up and the authorities had no difficulty keeping them from interfering with hunters' access to the area. Early on, three protesters from CEASE were allowed to pitch a tent in a circumscribed area adjacent to one of the main access points to the reservation, where they propped up several signs denouncing hunting, hunters, and the eating of meat. They vowed to remain encamped for the duration of the hunt.

There were no dramatic confrontations between defenders of animals and the hunters, as had been feared, and the TV crews had to content themselves with interviews of MDC officials whose early apprehensions began to melt away as it became apparent that no ugly scenes would occur. By midmorning, MDC spokespersons were clearly both relieved and pleased. Not only had protests and confrontations been averted, no accidental shootings had been reported. No hunters had been stricken by heart attacks (overexertion by hunters who are in poor physical condition is fairly common), nor had anyone required emergency care from the crews that had been brought in to respond to injuries sustained from a fall or to some other medically urgent need. The only remaining uncertainty, so far as safety was concerned, was whether any hunters had got lost. By dusk, all hunters had reported back to the checkpoints where, before dawn, they had been issued passes. No one had to be rescued.

Hunters were required to check their deer at stations inside the reservation manned by staff from the Division of Fisheries and Wildlife. In one of several departures from ordinary hunting regulations, hunters were instructed not to display their deer after they had been checked and tagged by division personnel, this to avoid inflaming public sentiment. As a result, when vehicles left the reservation it was nearly impossible for anyone, including the MDC officials themselves, to get a sense of how successful those hunters had been. By the end of the day, though, it became clear that the hunters had been very successful. As compared to normal hunts, where less than ten percent of all hunters bag a deer over the entire nine-day season, the 300 Quabbin hunters had killed 123 deer on the first day.

Upon learning of the tally, the three protesters from CEASE called for an immediate halt to the hunt. Using experience from other controlled hunts, the MDC had publicly estimated that roughly 170 deer would be killed in the course of the nine-day hunt. The protesters reasoned that since the kill on the first day alone had nearly equaled that targeted number, extending the hunt was unnecessary. Clifton Read, the MDC's interpretative naturalist and de facto public relations person, explained to the press that the estimate of 170 deer killed was not a goal, it was merely a guess as to what might result from the hunt. The goal was to sharply reduce the number of deer on the reservation as quickly and safely as was humanly possible, and the unexpectedly high kill on the first day, with no injuries to humans, indicated that the hunt was proceeding better than the MDC and Division of Fisheries and Wildlife had hoped. The hunt would continue as planned.

Rebuffed, the three protesters dug in for the long haul. The hunters' luck proved to be better than theirs. Weather conditions over the course of the nine-day season were nearly ideal for hunters—below normal cold and just enough snow to afford good tracking but not so much as to make walking difficult. There was some freezing rain, which didn't help the hunters, but it was not nearly as bad as what storms can bring in early December. Ideal deer hunting conditions, though, are not ideal protesting conditions. Though the three protesters persevered, their privations did not inspire others to join them. They hung on grimly, and with each day's tally they repeated their call for an end to the hunt. Midway through the hunt, one of the vigilers donned a hat decorated with fake antlers and, in defiance of MDC regulations, entered the reservation. The protester was promptly arrested for trespass and escorted out of the area.

By the end of the ninth day of hunting, the 900 hunters who had been randomly selected from the pool of applicants had killed 576 deer, far more than anyone had imagined would be taken. MDC and Division of Fisheries and Wildlife officials declared the hunt a success. Though the kill rate dropped off after the first day, even on the last day of the hunt, success in the Quabbin was higher than anything hunters achieve statewide. Since the hunters invited onto the reservation were not selected for their knowledge of woodcraft or their marksmanship, the high kill could only be regarded as dramatic confirmation of the MDC's claim that there was a superabundance of deer on the reservation. Hunters leaving the reservation confirmed that the number of deer was high. "There are deer all over in there," one hunter observed to reporters after the first day of the hunt (*Union-News*, 3 Dec. 1991, p.1).

Clifton Read speculated with reporters that the high kill suggested that there may well have been more deer in the area than the MDC had estimated.

The hunt was also successful in terms of logistics. The hunters who had been picked to participate in the "controlled hunt" had been required to attend one of ten evening training sessions that the MDC held in the weeks preceding the start of hunting. There they were told their assigned days of hunting and the gate they were to check in and out of on each of the days they hunted. They were also carefully instructed about the importance of using the portable toilets the MDC had deployed throughout the area to be hunted. And to further reduce the chance of contaminants entering the reservoir, hunters were told not to clean their deer near streams or wetlands. They were given a packet containing rubber gloves and towlettes so that they would not have to wash up in a stream after gutting their deer. Finally, they were admonished to stay well clear of the shoreline in order to avoid disturbing the eagles that perched along the water's edge.

In the post-hunt canvass of the area hunted, MDC officials were pleased to find almost no evidence of carelessness or rule violation. Moreover, a careful search netted only two carcasses of deer that, apparently, had been wounded but not retrieved by the hunters. (Crippling was a concern because of the lawsuit CEASE had filed in its effort to stop the hunt.) The high kill meant that the deer herd might be brought down to acceptable numbers much more rapidly than the MDC had thought possible and that the need for such hunts would not stretch out for years and years, as many critics and supporters of the hunt had feared. All in all, then, the MDC had reason to be satisfied with the outcome.

But opponents of the hunt were aghast at the scale of the "slaughter." Ray Asselin, of the Quabbin Protective Alliance, called the hunt "basically a massacre" (*Union-News*, 12 Dec. 1991, p. 26). The unexpectedly high kill led Steven M. Wise, president of CEASE, to claim that hunters had "wiped out the deer population" (*Union-News*, 21 Dec. 1991, p. 12). A few days before Christmas—just two weeks after the deer season closed—fifty demonstrators assembled at dusk on the Belchertown Common to hear speakers deplore the violence of hunting and to light 576 candles to commemorate the deer killed in the hunt. One of the demonstrators, Mark Picard, was quoted as saying, "Two weeks ago I lost 576 of my closest friends there." Wise, in an op-ed column in the *Springfield Union-News* (29 Dec. 1991, B3), condemned the hunt and called upon his fellow citizens to rebel against such violence and slaughter in the spirit of Daniel Shays. That Shays and his followers intended to shoot their way to independence from Boston financiers was somehow overlooked by Wise. Clearly, the passions aroused in the months leading up to the hunt were not going to dissipate. The hunt would continue and so would the outcry.*

Living with Nature

Nature might well be thought of as the original Rorschach. Like the suggestive, amorphous ink blots psychologists use to tap our innermost fears and longings, nature presents an open invitation to see what we want or need to see. The woods that seem dark and foreboding to one person appear the model of tranquillity and harmony to another. This is as true for those who bring the full power of science to their understanding of nature as it is for amateur nature lovers; for the most ardent and experienced backpacker who is regularly immersed in wilderness as for the devoted couch potato whose closest encounter with nature is a Disney rendering of the "wild kingdom." Each of us has a version, a set of beliefs about nature. Some versions fit more closely with the one that is commonly accepted by experts as "true," but the truth of one or another version is, after all, a matter of convention—what others agree to ratify as "reality."

This is not a comforting notion to accept. We would prefer, I suspect, detective Jack Friday's view of truth: "just the facts, ma'm." But facts are slippery things. They change with each interpretation, and

---

*In the late summer of 1992, CEASE reintroduced its suit to stop the hunt. Again, the courts dismissed their arguments. Hunting was expanded to a new, heavily impacted area, the Prescott Peninsula, and was resumed in the area of the 1991 hunt. The kill on the Prescott was high, in effect a repeat of the previous year's experience. In the area first hunted, the kill was much lower, but still somewhat higher than would be expected in comparable areas off the reservation.

facts that at one moment seem decisive can be upstaged abruptly when something unexpected occurs. It is, then, small wonder that our ideas about nature are continually in flux, continually subject to rejection and radical reinterpretation.

In the face of this inherent ambiguity, humans have created intricate cosmologies to explain nature and natural phenomena—the movement of the stars, the passage of seasons, the awesome power of an electrical storm. Hunter-gatherers developed complex schemes to account for the medicinal properties of plants and animals. Elaborated accounts of causal interactions between natural forces fill the storehouse of human cultural achievement. Everywhere and at all times, people have struggled to understand themselves in relationship to nature.

This curiosity runs to the very heart of our identity. We are a boundary-drawing species. Whether the boundaries distinguish between groups of humans, or separate humans from other species, or involve such complex taxonomies as the periodic chart of elements or the Linnaean classification of the world's flora and fauna, humans have labored to draw lines and to order things on the basis of some range of real or imagined properties.

Naming things and thereby placing them into a scheme is to assert power over the things named. It may also be reassuring, insofar as it helps to reduce the mystery of what moves the world around us. We do not have to subscribe fully to the anthropologist Bronislaw Malinowski's theory that religion and magic originate in the attempt to reduce anxiety in the face of overwhelming uncertainty in order to appreciate the comfort that can be drawn from knowing that there is some order and purpose to things.

Much, if not all, of human culture—including the middens, temples, amphitheaters, office parks, and domed stadia—is an expression of our fundamental insecurity, our deep sense of fragility and impermanence as we confront the natural world. Hunter-gatherers were clearly closest to being at home in the wild, judging by how little they seem to have attempted to alter their environment. To be sure, they developed complex pharmacopeia with which they healed themselves. And they devised all manner of gods and spirits whom they propitiated in order to gain their protection and solicitude. Still, they left little imprint so far as we can tell, and in this respect hunter-gatherers are the exception. The rest of humankind has endeavored to modify nature, however modestly. Whether this entailed setting fires to aid in clearing land or to drive game into ambush, or diverting watercourses to irrigate fields or to power gears, or deliberately breeding plants and animals to achieve a desired end, humans have been busy for thousands of years tinkering and tampering in order to control and shape nature.

There have been some truly remarkable successes, successes that in the aggregate and over a very long stretch of time have meant that human beings are both more long-lived and far more numerous than would otherwise be the case. But it is hard to make the claim that we are "better off" in other respects. I am inclined to say that we are, that life, even for people in remote areas of the earth, is richer and better now than it was, say ten thousand or even one hundred years ago. Nevertheless, I accept the very deep ambiguity attached to such a claim. Part of the reason for this ambiguity is that the record of interventions is mixed, to put it mildly. There have been calamitous failures, efforts that went bad, not only failing in their own terms but actually making matters worse. A list of these follies could easily fill a large volume.

If such a list were constructed chronologically the bulk would no doubt swell as we near our own time. But the outrage over ozone depletion, toxic waste, and desertification, to take only three examples, should not blind us to the ways that our ancestors messed up. The famed biologist René Dubos has commented that many of the places we now celebrate for their stark beauty, places like Sardinia and Greece, are in fact environmental disasters created millennia ago by heedless logging and equally shortsighted grazing. But these things happened so long ago that we now associate the landscapes of the Mediterranean with the beauty of antiquity, not with environmental degradation. If there ever is a final accounting, Homo sapiens will have much to atone for. Specific failures are chastening, but the record of failure does not seem to have greatly diminished the enthusiasm with which humans have sculpted, engineered, and otherwise modified nature. A mess seems to spur us on to redoubled effort, not to a reappraisal of the enterprise itself.*

This willingness to proceed, despite the record of mistakes and the risk of calamity, stems at least to some degree from the basic uncertainty we feel about untamed nature. In a very real sense, we are not at home in our landscape, however familiar we may be with it and whatever the affection we have for certain of its features. Some of this insecurity may stem from the sense of sheer puniness we all feel in the presence of a storm or of a majestic mountain peak. Our most complex designs are easily surpassed by the intricacies of a spider's web or the spectacular detail of a butterfly's wing. But more than awe is involved.

Nature, as I have noted, is contested terrain. Not only are humans struggling to wrest a livelihood from nature, we are also struggling with one another over the very definition of nature's meaning. The sheer complexity of interactions that must take place to produce a change in seasons, an upswing in the squirrel population, or a drought leaves much room for debate, skepticism, or awe. Interpretations change, often radically, with new discoveries. Our sense of what is "out there" is variable, in part because our knowledge is at best partial. Also contributing to this variability is the fact that nature is itself changing, however imperceptibly.

Early hominids must have experienced themselves as scarcely separate from the other life forms around them; like everything else, they ate and were eaten in turn in what must have seemed an endless cycle. The dawn of civilization is marked by the assertion of separateness and by attempts to affect the cycle—to prolong life, to gain a competitive advantage over other predators, to take some measure of control over

---

*Granted that there is a long history of misgivings, of people resisting attempts to alter this or that feature of the environment. Interpretation of the biblical injunction to claim dominion over all of creation has varied considerably—from early expressions of what we would recognize as "stewardship" to unabashed enthusiasm for domesticating nature. Animal rights advocate, Lewis G. Regenstein, while acknowledging that the latter has been the dominant reading of the Judeo-Christian tradition, is at pains to note that there is a basis for animal rights in the discourses of theologians in ancient as well as modern times. But it is clear that misgivings and resistance grow steadily more notable and insistent the nearer we come to the present. See Regenstein's *Replenish the Earth: A History of Organized Religion's Treatment of Animals & Nature—Including the Bible's Message of Conservation & Kindness to Animals*, New York: Crossroad (1991).

human fertility. Separateness, of course, raises the question of standing: are we "separate but equal" or are we superior? Should we fit in, adapting ourselves to the cycles of feast and famine as they come at us? Or should we endeavor to regulate and modulate the ups and downs?

For better or worse, the matter is settled in all but ideological ways. We may wish to be equal, to be one with nature, but it is hard to imagine how we could possibly break the cycle of management except in a few out-of-the-way preserves. And even these can be kept secure only if they happen to be found to contain nothing of value. As it turns out, however, simply being left alone produces something of value: "wilderness." People will flock to see what it is like, and unless we absolutely ban all human access to such areas, they too will have to be managed sooner or later. But if we must ban all human presence in order to keep something "natural," we are declaring that humans are "unnatural," that we do not belong. Surely this is a curious turn of logic.

We are trapped. In this sense, McKibben is right to announce "the end of nature," though his timing is off considerably. If we mean by "nature" that which is unaffected by human action, nature ended quite a while ago, back when humans began deliberately to alter their landscape to suit them. Daniel Hillel's book *Out of the Earth* (1991) makes this point emphatically. A soil scientist at the University of Massachusetts, Hillel shows that even the humble efforts of early agrarians transformed forever the character of the soils, and that with this alteration, the nature of the plants and animals that could thrive on the land was also forever altered. No doubt, our chemical and radioactive wastes represent both greater and more dangerous changes than our ancestors' interventions. If only for this reason, McKibben is correct in urging us to face the apocalyptic consequences of our disregard for the environment. But that is a separate matter. On the issue of nature existing as a force unaffected by human artifice, it is clear that that kind of nature disappeared millennia ago. We will never get it back.

If "nature" in this narrow sense ended long ago, our sense of innocence about nature persists. And that's the rub. Our forebears can be forgiven their innocence. Their tamperings were, for the most part, modest and the consequences were generally imperceptible save over long spans of time far exceeding individual and in many cases even collective memory. But we cannot be so easily forgiven our naiveté. To pretend that things can be restored to their former condition and then kept that way, especially by benign neglect, is a dangerous folly.

Even were it not a reservoir, the Quabbin could not be restored to its original condition, whatever that may have been. Moreover, even if we could somehow achieve such a restoration, it is impossible to imagine how we could maintain it in that state of suspended animation without herculean effort. To take but one aspect of the problem, we would have to somehow reintroduce large predators, notably mountain lions and wolves, in order to contain the deer herd. This would require that residents for miles on all sides of the reserve tolerate the food preferences of these creatures. Large predators cannot be trained to stick to deer. With comparatively easier food available in domestic herds and flocks, not to mention pets, they would seem to be more likely to specialize in these morsels than in the harder-to-catch deer.

This is an old problem, one that Leopold grasped over fifty years ago. Certain features of the wild are enormously compelling; unfortunately, they come with other features that are not nearly as compelling nor as easy to abide. Deer have been among the more compelling to our sensibilities. They have managed to capture the hearts of hunters and antihunters alike. Whether as game animal, food, or symbol of pastoral tranquillity, we have encouraged deer to propagate. At the same time, we have reduced the traditional predators of the deer to remnant populations. The result: everywhere deer herds are booming.

It stretches the point only a little to say that the Lower Forty-eight is becoming a giant Quabbin. Scrawny, emaciated deer browse day and night across the El Cerrito-Berkeley-Oakland hills, their foraging adding to the instability of the slopes by diminishing the ground cover. The number of deer-vehicle accidents mounts and the incidence of Lyme disease rises. The Quabbin debate, with necessary adjustments for local conditions, is being played out in Princeton, New Jersey, Fire Island, New York, and Philadelphia, to name but a few of the places where deer have become a problem for public safety and health.

Similar stories can be told of the Canada goose. A succession of mild winters in the Northeast, coupled with new food sources made available by suburban sprawl, has caused a larger and larger number of geese to spurn the rigors of flight to their traditional wintering grounds in the South. Year-round populations of geese now pose a nuisance to park and golf course managers. More important, they are beginning to threaten the safety of some municipal water supplies.*

Seals, too, are becoming a problem. Environmentalists and animal rights advocates essentially put an end to the hunting of seals nearly twenty years ago. Spared the clubs of men seeking the skins of baby seals, the population of seals has exploded to the point where they have begun to compete seriously with humans over the already stressed and overharvested fishing grounds of the North Atlantic. In the spring of 1992, marine biologists fretted over the possibility that the exploding population of seals would invite a viral epidemic that would devastate the seal herd. Are the added stresses on the fishery or a widespread die-off of seals preferable to sealskin coats?

Choices such as these are morally troubling. It would be comforting to be freed of the burden of having to make them. Those who hold that nature, left to itself, balances out deer, geese, and seals with supplies of fish, trees, water, and bacteria seem to occupy a moral high ground. But it is a high ground of their own creation, fabricated out of a particular set of beliefs about nature and its processes. Theirs is a morally simplified world, like the world of the laissez-faire economist. It is a world freed of the troubling need to take responsibility for outcomes.

Unfortunately, this is not the world we actually inhabit. The world we inhabit is a world we have had a large hand in making. Not a free hand, mind you, just a large hand. We have tried, to paraphrase Marx, to make the world just as we please, and we have repeatedly failed, sometimes spectacularly. But having tried, we cannot now walk away from this history. We cannot start afresh. There are no clean slates. We have, I am afraid, no alternative but to face squarely the need to manage and to accept the burden of making choices.

*The managers of the Quabbin have begun to worry about their goose population. For now, they are hoping that by planting ground cover that the geese do not like around the outflow areas of the reservoir, they can keep the coliform bacteria count well below troublesome levels. But if the goose population continues to rise, goose hunters in camouflage may join the deer hunters in blaze orange on the reservation.

Living with Ourselves

There is no impact-free way for us to exist. Some things, indeed lots of things, must yield, adjust to our actions and the conditions we create—or perish. Over time, we have grown more conscious of the ways we impact upon nature, becoming steadily more deliberate in our attempts to manage and manipulate the natural order, and more aware of the effects our activities have had on the environment. This has meant that the sense of burden, even guilt, has also risen. For some, this sense of burden drives a search for an optimal number of humans, or an ideal life style that, if adhered to, would permit us to coexist harmoniously with the rest of creation. Many of the critics of the MDC yearn for just such an ideal state and are actively trying, in their own personal lives, to live as simply, as modestly, as they can in order to reduce as much as possible the harm they do to nature.

However admirable and beneficial these attempts are, they do not address the fundamental problem: we must extract resources from the earth and, in the process, alter nature. This is as true for humans as it is for deer. We cannot say what the ideal number of deer or humans is because the answer depends on what we think the entire ecosystem should be like. Our answers will be profoundly subjective ones, a function of values and ideals about which there is no general agreement. In the end, arguments about nature are really veiled arguments about human values and the social arrangements to which they give rise.

Environmentalism is inseparable from moral reform, though the language of biology and chemistry in which discussions of the environment are often couched tends to mask this connection. Thus, when the ecologist David Ehrenfeld urges his readers to study and emulate nature, he does so thinking that these activities will produce humans who will be less inclined to ravage and plunder heedlessly. Why? Because nature teaches "honesty, reliability, durability, beauty, even humor" (ix). Does it really? Doesn't nature equally teach "eat or be eaten," "survival of the fittest," or any number of other lessons that would seem to authorize the very arrogance and selfishness that Ehrenfeld so rightly deplores?

In the same vein, we have seen that much antihunting sentiment arises from the conviction that if hunting were ended, people in general would be less violent and abusive, not only toward animals but toward their fellow humans as well. Similarly, solicitude for animals is thought to produce better people. The same sort of arguments have long been made in behalf of vegetarianism. Turning away from meat will, it is held, make us more pacific. The evidence for such assertions as these is slender, verging on nil. The absence of hard evidence, though, does not diminish the tenacity with which such convictions are held, any more than the ecologists' rejection of equilibrium models reduces people's attachment to the idea that nature represents balance and harmony.

Paul Lyons, the MDC's wildlife biologist and one of the few in the MDC faction who had some sympathy for the philosophical grant of rights to nature, seems to me to have got it just right when, echoing Leopold's letter to the managers of Glacier National Park, he said that we must stop talking abstractly about nature and natural balance and instead focus on determining how we want to shape our environment. Let us be explicit about what we value and why, and then face up to what these preferences require of us. If we want less management, fine. But let us not pretend that we will have to make no adjustments in the way we live. If the adjustments do not seem warranted by the gains, then we will have to accept the need for more management and manipulation of nature.

As I have suggested, most of the people who were at loggerheads over the deer hunt would agree that in general we should reduce the stress we are imposing on the environment. As consumers, we should make do with less. We should be much more willing to abide the randomness of nature—let streams meander, let yields vary, let patches, the larger the better, "go wild." Why, then, was the division so deep and so bitter? Why, with so much in common, did the prospect of a "deer reduction program" become so inflammatory?

Part of the answer, as we have seen, is the low regard in which hunting and hunters are held by those who are, broadly speaking, environmentalists. The harvest mentality strikes many as, at best, inappropriate, and at worst as murderous. Many environmentalists also deeply resent the pressure hunters exert to keep lands open and accessible. They would rather have as much land set aside as possible and hold access to a minimum in order to preserve what is left of unspoiled or recoverable natural areas. Evironmentalists all too regularly encounter hunters as opponents, allied with interest groups that are pressing for intensified use of the environment. Hunters may see themselves as custodians of nature. Most environmentalists see them as foxes guarding the chickens.

But more than antagonism to hunting and the political alliances hunters have entered into was involved in the partisanship revealed in the dispute over the deer of the Quabbin. If my reading of it is anywhere near the mark, the division is a continuation the longstanding split between preservationists and conservationists that Leopold tried so hard to transcend. Are we to be active caretakers, in the process imposing our values and ideals on the canvas nature presents? Or should we be passive caretakers, wherever possible letting nature take its course?

These two postures would not be as enduring as they have been were there not good and compelling arguments for them both. It is easy to lose sight of this truth in the heat of passionate controversy such as enveloped the managers of the Quabbin. The critics, in the final struggle, could only see the MDC side as resource exploiters, an unholy alliance of men wielding chain saws with men brandishing shotguns. For their part, the supporters of the MDC saw their critics as unreasonable, if not irrational, idealists whose heads were filled with fanciful notions about forests and deer. These caricatures, of course, don't come close to describing the partisans on either side, and not simply because the members of each camp were diverse and arrived at their respective positions in complex and thoughtful ways that the caricatures do not begin to capture. More important, the caricatures fail because each denies legitimacy to the other side.

There is a sound argument for letting the Quabbin alone, and shifting resources from the management of the watershed to efforts to decrease consumption and waste of water. If substantial reductions in consumption were achieved, worries about water yield, forest cover, and deer would fade and the "accidental wilderness" could continue to change according to its own lights. Whether balance and harmony or chaotic fluctuations would result would matter little since, with moderated consumption, the margin for error would be comfortably large.

The proponents of this view have a faith in nature's ability to provide—if we can discipline ourselves to control population growth, minimize waste, and stop equating progress with ever higher levels of personal consumption and more intensive utilization of natural resources.

There is also a compelling argument for keeping a steady hand on the helm at the Quabbin. Nature is, as we know, full of surprises. The Quabbin is a hedge against surprises. The intervention that created the reservoir makes subsequent management of the watershed look miniscule by comparison. With so massive an intervention to start with, and with so many lives and livelihoods now dependent upon this man-made source of water, it seems only sensible to take measures that will keep the watershed and reservoir buffered against surprises and as resilient as possible.

Since no one in his or her right mind would now propose to drain the reservoir, we have to own up to the disruption we have caused and the resultant responsibility to manage it. If deer or beaver or any other species, animal or vegetable, begins to crowd out other species or otherwise pose a threat to the long-term stability of the watershed, intervention designed to keep things more or less stable will have to occur. This will mean chain saws and guns.

It does not mean we have to accept rude and menacing ruffians roaming the countryside any more than we have to accept clear-cut moonscapes in the wake of timber harvests. But it does mean that we have to accept the necessity of activities like hunting and logging that many find morally objectionable, because these activities, closely monitored and regulated, are crucial to the maintenance of habitat and animal populations.

How are we to choose between these competing arguments? Is there a golden mean, some common meeting ground where the opposing views can be reconciled? I am not nearly as optimistic about the prospects of such a reconciliation as I was when I embarked upon this project. The problem is not the size of the gulf separating the two sides. As we have seen, on most points there was more often agreement than not. Everyone, even the most enthusiastic and unabashed boosters of management among the hunters, acknowledged the urgent need to find a more sustainable relationship between humans and the environment. Almost everyone agreed that we have to control human population growth. What was beyond reach, however, was an agreement on the meaning of nature.

Deer and oak trees were but the visible markers delineating the battle lines. The real fight was over symbolic turf. People had projected onto the Quabbin the qualities they longed for nature to possess. For people who believe deeply, *and with good reason*, that "nature knows best," one more turn of the interventionist screw, especially one that threatens to open a reserve to the violence of hunting, is sacrilege. Animal rights advocates may fear the loss of a few hundred deer, but the bulk of the critics feared a far larger loss—the loss of one more place where an embattled vision of harmony and unity had stood a chance and might have become a beacon. There is no possibility of compromise, and no consolation, when the stakes are that high.

The stakes for the MDC faction were not nearly so high. I do not say this to suggest that the critics were inflexible or unreasoning and the MDC faction, by contrast pliable and open. Both characterizations would be badly misleading. Nor is it the case the the MDC faction had a more "realistic" view of nature. Its view was fully as much a *belief about the nature of nature* as was the view of its critics. Rather, because the MDC faction tended to see nature in more instrumental terms, or, if you wish, in less transcendent terms, it could more easily maintain a willingness to compromise. Members of the faction freely accepted the notion that management practices and goals change, sometimes in response to altered natural conditions, as with drought, sometimes in response to changing political winds, as with the lobby for fishing.

Although they shared their critics' dismay at our wasteful ways, members of the MDC faction could not as readily envision humans living in harmony with the land. We are a dominant in the order of things and, like other dominants, we willy nilly leave our imprint. Better that we be self-conscious and planful about the imprints we leave. The MDC faction was convinced of the need for management, though it was not committed to any particular management policy. Its members could not, finally, share the idealism of many of their critics. Though they were, hunters aside, no more enamored with hunting than their opponents were, they could not believe that abolishing hunting would speed the arrival of civic harmony and peace, or that it would allow the deer and the antelope to play. Both humans and animals are more complicated than that.

Because of these differences, even though the MDC's position permitted negotiation and the possibility of compromise, each side framed the debate in ways that virtually precluded accord. When the MDC, in the final plan of which the hunt was the major element, agreed to set aside several thousand acres to be left unmanaged, the critics were not in the least mollified. Some saw the gesture more as an insult, a trivialization of their concerns, than a conciliatory offer.

The collison over the management of the Quabbin foreshadows a sharper and more divisive debate over environmental policy and natural resource management that we have periodically joined but about which, as a nation, we have not yet gotten serious. As pressures on the environment mount, the policy decisions will grow steadily more agonizing and become less a choice between competing goods than a choice between evils. In many cases the differences between evils will be so slight that we will be deprived of even the cold comfort of choosing the lesser. It is not a pleasant prospect. If we cannot agree on the nature of nature and accept ourselves as responsible shapers of what we want nature to be, we are not likely ever to agree on how we ought to live with nature.

I drove by the Quabbin just the other day, a cold, clear winter's day. The water was calm and brilliant blue. The leafless trees were dormant. The echoes of the fall hunt had long since died away. The woods looked peaceful. It was hard to believe that this could have been the backdrop to such a bitter dispute among people who, in all sincerity, professed such deep and abiding love for the place. It is sad, grievously sad, that the people who love the land, albeit in different ways and with different needs, cannot unite and agree how best to defend it. If the people who think and care most deeply do not make common cause, there will be precious little that future generations will be able to call "accidental wildernesses" and mean by that something inspiring, mysterious, and comforting. It will be an unimaginable loss.

7

Stewards or Curators?
Caring for Nature

The idea of "letting nature be nature" arises . . . from sec-
ondhand knowledge and nature-romanticism; it does not
work in practice.

MARY ZEISS STANGE

NATURE IS ALIVE on so many scales, both spatial and temporal, that
it is hard to be certain that we ever really know what is going on.
Models generated from close study of one setting do not necessarily
travel well. Subtle variations in climate, soils, precipitation, and a host
of other variables can make for decisive differences in the interactions
of flora and fauna. In all this welter, there are few general principles on
which to base predictions. One of these is so reliable as to be very nearly
a cliché: nature abhors a vacuum. Like most clichés, it contains an
important truth: nature is constantly in flux and change itself becomes
the opportunity for more change. The MDC created a vacuum when it
cleared the Swift River Valley of buildings and wood lots in the 1930s.
As we saw, nature began filling this vacuum with what would have been
a more or less typical New England forest cover—species of oak, maple,
cherry, ash, and, of course, birch, hemlock, and white pine, to name the
dominant varieties. These also happen to be the ingredients for gour-
met dining for white-tailed deer. Absent predation, including human
predation, the deer began creating their own vacuum. Their food preferences began to dictate what would and would not grow.

For reasons we need not rehearse, in the late 1980s the MDC set a course to reduce significantly the size of the deer herd by means of a controlled hunt, in effect creating another sort of vacuum into which, it hoped, a wider range of tree species and age classes would grow. When the controversy over forestry practices and the hunt first flared, the intensity of the conflict obscured an important feature of the MDC management plan. The MDC wanted a forest with virtually all the qualities that its critics valued—a forest more nearly like what would have been there had no reservoir been created. The problem was not simply that the deer would not let such a forest develop on its own, but also that the deer stymied the MDC's efforts to help such a forest develop. Well established silvaculture techniques, instead of encouraging regeneration and species diversity, simply generated more deer browse. To make matters worse, chronic overbrowsing had produced a forest that, even if left to itself in the absence of deer, would not yield much regeneration or diversity until some major disturbance chanced along. The combination of a thick canopy of mature trees and an equally thick mat of ferns on the ground did not make for favorable tree growing conditions.

The result was deeply paradoxical. Everyone wanted to see the Quabbin forest become richly diverse, resembling as closely as possible a "typical New England eco-system." Given the deer, however, the only way to get to that "so-called natural" condition was to manage the deer herd and the forest, keeping the reproduction of the former down while boosting the regeneration of the latter. Advocates of "nature's way" refused to accept the paradox. Managing nature was anathema, period. With the hunt a fait accompli, the MDC would have their practices put to the test.

The success of the first hunt in 1991 was repeated in 1992. Though fewer deer were killed in Pelham the second year, the kill on the newly opened Prescott Peninsula was even higher than it had been the first year in Pelham. In all, 724 deer were killed in 1992, compared to the first year kill of 576. Again, there were no accidents, no dead eagles, and this time there was only a small symbolic protest by a handful of animal rights advocates. With the hunt on track, the MDC began to plan its strategy for the forest's recovery. In 1994, the first of what would become an annual public Quabbin Land Management Workshop was held to inform all interested parties about what was being observed on the reservation and to present a draft of what was ultimately to become the "1995–2004 Quabbin Land Management Plan." The draft plan called for a pattern of selective cuts involving roughly five hundred to fifteen hundred acres each year. Each of the cuts would be quite small, few exceeding five contiguous acres, and they would be spread out over the fifty-five thousand acres of the reservation. Some of the cuts were designed to initiate regeneration essentially by clear cutting (in the parlance of forestry, these are called "regeneration cuts"). Where desired seed stock was absent, as in many of the red-pine plantations, seedlings of the desired species would be planted in these cleared areas. Otherwise, the clearings would be left to fill in by themselves. Once regeneration was underway, the MDC would conduct "release cuts": the selective removal of mature trees around regeneration plots so as to allow more sun to spur the young trees' growth. These areas would be closely monitored to assess the extent of deer browse. In addition, the MDC installed electric fenced exclosures, mostly in the Quabbin Park, the only area of the reservation not slated to be hunted.*

With the regeneration and release cuts underway, additional sections of the reservation were opened to hunters. In 1993 Hardwick and New Salem were hunted for the first time. Considerably fewer deer were shot in these two areas, compared to the first-year kills in Pelham and on the Prescott. Still, with four of the six areas of the reservation now being hunted, 474 deer were killed in 1993. In 1994 Petersham, the last of the areas to be hunted, was opened and the number of deer killed rose sharply to 673 (456 in Petersham alone). In 1995 the number of deer killed dropped sharply to 284, raising hopes that control over the deer herd was at hand. The 1996 season was a poor one for hunting. The weather was miserable, and with each area hunted for only three days, many hunters had their entire three-day hunt washed out. Only 129 deer were taken. In 1997, with more favorable weather, the number of

---

*The so-called "park" is the only area of the reservation to which the public has open access during daylight hours. The roads are hard surfaced and there are pull-offs as well as an observation tower that afford panoramic views of the reservoir and surrounding forest as well as picnic areas. It is in this section of the reservation that visitors used to feed the deer.

deer taken doubled the previous year's kill but was still far below the peak of 1992, when only two areas were hunted. The declining harvest appeared to signal a significantly reduced deer herd.

It took only a few years for the results of this two-pronged strategy of stimulating regeneration and keeping the pressure on deer to be noticeable. Young saplings began to appear and, most crucially, many were escaping the attention of deer. To be sure, regeneration was still more robust within the fenced areas than on the outside, suggesting that the deer were continuing to have an impact on the forest. By the 1996 Quabbin Land Management Workshop, with the last of the five years of hunting fast approaching, more definitive results began to come in. A botanist from the University of Massachusetts who had been studying the Quabbin for years reported that she and her graduate students had found a few wildflowers, the first that had seen on the reservation in memory. To this encouraging news was added more precise information regarding tree regeneration. It is worth quoting at some length the 1996 report of the MDC Land Management Program.

> The results of regeneration monitoring continue to be encouraging, although they are quite variable by area. For the watershed reservation (previously not hunted) as a whole, mean per acre regeneration values for "disturbed" areas (that is, where light is not limiting and regeneration is expected to occur) have progressed steadily. In 1989, these means showed a total average number of regeneration stems (one foot tall to one inch in diameter at breast height) of 1533. This same summary averaged 3237 in 1994 and 3629 in 1995. Likewise, there has been steady progress in both the unfenced and the fenced permanent $1/100$th acre regeneration plots in Pelham. The mean value for regeneration $1'$ to $4.5'$ tall for unfenced plots has moved steadily from 38 stems per acre in 1991 to 208 stems in 1995; for regeneration greater that $4.5'$ tall the numbers have moved from 3 to more than 20 per acre. For fenced plots, regeneration from $1'$ to $4.5'$ has moved from 47 stems in 1991 to 708 in 1995; for regeneration greater than $4.5'$, the numbers have moved from 0 stems, on average, to 114 per acre in 1995.

Progress was real and measurable, but these figures also made plain that it would be some time before the effects of decades of deer browse would no longer be apparent. The report blandly noted that "plans for

FY-97 call for a continuation of controlled hunts in all five areas. . . . "
The initial proposal over which the controversy raged in the early '90s
was for a five-year hunt. Some of the original opponents of the hunt
had feared that the five-year hunt would simply be an opening wedge
and that hunting would, in fact, become an institutionalized feature of
the management of the Quabbin. They were right, though for wrong
reasons. Opponents of the hunt assumed that once in, hunters would
form a strong lobby and, with the support of the Division of Fisheries
and Wildlife, simply muscle the door to hunting wide open. In fact, as
we shall see in greater detail below, hunter interest in the Quabbin hunt
would drop quite precipitously after the 1996 hunt, and the Division of
Fisheries and Wildlife has become more a pliant supporter of the
MDC than the juggernaut the antihunting forces feared. Regeneration, not a hunting lobby, drove the decision to seek an open-ended
continuation of the hunt.

The problem the MDC faced in 1997 was the result of the success of
their efforts to spur tree regeneration. Regeneration had begun, but as
the numbers cited above make clear, the bulk of this new growth was
under 4.5 feet, the height generally regarded as safe from significant
deer browse. In other words, to stop hunting deer, having just produced
a crop of high quality deer habitat, would mean only one thing—a surge
in deer numbers. The fecundity of deer rivals the legendary capacities
of rabbits. One of the longest running and most carefully controlled
studies of deer showed that in good habitat, such as the Quabbin, deer
can attain a herd growth rate of 50 percent per year if hunting is taken
out of the equation.[1] In other words, ceasing the hunt after the fifth
year would, in all likelihood, put things right back to where they had
been in 1990.

Another problem was equally clear by 1996: despite the evident successes of the hunt, regeneration in most areas of the reservation was still
markedly lower than it was in similar areas outside the reservation.
Perhaps deer reproduction was keeping pace with the hunters, with the
net result less reduction in the size of the herd than the kill was leading
everyone to believe. By the 1998 workshop, held in early June, the
question of the deer herd size became all the more pressing. In all but
one of the five areas of the Quabbin, results of deer browse surveys,

while showing a downward trend, were still hovering at or very near the mark foresters considered to be "severe": when 30 percent of all available stems in test plots show evidence of having been newly browsed. For example, in Pelham, the area hunted longest, the browsing survey showed a 28 percent level of deer browse. Two of the areas, Prescott and Petersham, were still at 30 percent. Only in Hardwick, where the numbers of deer killed had not even been as large as in most other areas, were the declines in browse dramatic, down from a whopping 49 percent in 1996 to a minuscule 3 percent in the spring of 1998. The 1998 browse survey turned up a new source of concern: "significant evidence of browsing at heights above 5 feet and up to eight [feet] or more." Moose.

When *Going Wild* went to press in 1993, there were occasional sightings of moose in the area. In 1990, when I interviewed Paul Lyons, the MDC's wildlife biologist, he was certain that moose had become regular visitors to the Quabbin but doubted that a permanent breeding population had taken up residence. Since then, things appear to have changed rather rapidly. The moose population has grown steadily, and by 1997 it had become large enough to spur the state highway department to erect signs warning motorists of moose crossings on the major east-west highway that runs just north of the Quabbin. It is still not absolutely clear that moose actually have set up housekeeping at the Quabbin, but if they have not, it is just a matter of time—and the time will be measured in a year or two, not decades. There can be little doubt that soon the MDC is going to have to confront the damage done by moose. But this gets us a bit ahead of the story. Let us return, briefly, to 1996–97.

As noted, the 1996 workshop report indicated plans for continuing the hunt. The biologists and foresters managing the Quabbin clearly regarded this extension an open and shut case. The data the MDC had gathered over the previous five years clearly vindicated the MDC position on the relationship between deer and forest regeneration. And despite the evident successes of the annual hunts, the continuing evidence that deer were continuing to depress regeneration was compelling reason to continue keeping pressure on the deer herd. There was also no doubt about the agency's ability to manage a controlled deer hunt effectively after five years of virtually uneventful hunting. In fact, the only blemish on the hunt came in the third year when a hunter mistook a female moose for a deer and killed it.* Moreover, public protests over the hunt had, by the third hunt, virtually disappeared, and the most vocal of the agency's critics, the Quabbin Protective Alliance, had dissolved, a fate common to single-issue grassroots protest groups. It is no wonder that the Quabbin managers took the reauthorization of the hunt for granted.

Behind the scenes, however, the two well-established organizations that had opposed the hunt from the outset, the MSPCA and Massachusetts Audubon (not to be confused with National Audubon), went directly to the MDC Board of Commissioners, a group of political appointees far removed from the Quabbin who are in charge of overseeing the operations of the sprawling state agency which not only looks after watersheds but also has skating rinks, parks, and other public assets spread far and wide across the eastern half of the Commonwealth. Much to the surprise and chagrin of the people in charge of the Quabbin, the board voted to suspend the hunt before even listening to presentations from their own staff. A hue and cry arose, and within days a chastened board solicited the views of the people directly in charge of managing the Quabbin. After hearing the careful analysis of the Quabbin managers, the board reversed itself and authorized the continuation of the hunt. As if it hadn't been made abundantly clear earlier, it was certainly now clear that the managers of the Quabbin would be harried both by ungulates and by organizations committed to letting nature take its own course.

The point was driven home in the spring of 1998: the Quabbin Watershed Advisory Committee (QWAC) recommended that hunting in New Salem be stopped "as an experiment," despite the unanimous recommendation of the Quabbin managers that the hunt go

*This, of course, was a clear violation of the state's game laws. The hunter reported his mistake immediately and was subsequently fined and had his hunting privileges withdrawn for several years. Because he was forthcoming, this punishment was considerably lighter than the law allowed.

forward on all five areas and despite the evidence of continuing significant deer browse in New Salem.* QWAC had evolved out of a group
made up exclusively of people interested in issues related to access to
sport fishing on the reservoir. With broader environmental issues rising
to the fore, the legislature revised both the membership and the charge
of the committee and gave it a new name. QWAC's membership continued to include delegates from the Commonwealth's fishing and
hunting communities but now also included members drawn from the
broader spectrum of environmentalists, including representatives from
Massachusetts Audubon, the MSPCA, the Sierra Club, as well as
people from the communities sharing common boundaries with MDC
watershed lands.

Massachusetts Audubon and the MSPCA, having lost their campaign to cancel the entire hunt the year before, decided to settle for a
fifth of a loaf. The two organizations argued that it was time to see if
the hunt was really a continuing necessity. New Salem seemed a good
place to start. The deer kill in New Salem had been consistently low
(only nine deer were shot in the New Salem area during the 1997
season), in part because the topography and forest cover there make it
much harder to hunt than the other sections of the reservation. Despite
the relatively low deer kill, regeneration in New Salem has been rather
impressive, suggesting that the herd there may have been smaller to
start with. The 1997 survey of test plots shows three times as many
stems per acre in New Salem as on the Prescott Peninsula. Indeed,
New Salem compared favorably to the number of stems per acre in test
plots off the reservation. The combination of a low deer kill and relatively high regeneration lent a superficial legitimacy to the proposal to
suspend hunting in New Salem. To everyone's surprise, the QWAC
members representing hunters sided with their old enemies, once more
confirming the old saw that politics makes for strange bedfellows. The
hunting members of QWAC no doubt reasoned that suspending the
hunt would result in a larger herd, which would compel the MDC to
reopen the area to hunters, who would be greeted, in turn, with good

---

*Though technically only advisory, the managers of the Quabbin have made a point of taking
QWAC very seriously, not least because they have a strong preference for avoiding confrontation
and the bad publicity that inevitably would follow.

hunting—that is, more "targets." The MDC personnel with whom I spoke were not surprised by the position taken by MSPCA and Mass Audubon, though their continued refusal to acknowledge either the initial problem or the incontrovertible evidence of the success of the MDC's strategy was a source of disappointment verging on bitterness. But, having received enormous abuse over hunting, the MDC was genuinely dismayed by hunters so quickly abandoning their initial professed willingness to be "tools of management." By breaking ranks with the MDC and its goal of managing a resource in an environmentally sensitive way, this group of hunters confirmed what antihunters and many in the MDC had thought all along: that hunters are interested in the environment only to the extent that this means more opportunities to hunt. The net result was that there was virtually no public support for the MDC's position that all five sections of the reservation continue to be open to regulated hunting.* QWAC bought the idea of the closure as an "experiment."

The managers with whom I spoke were dismissive of this "experiment." From their point of view, there was no need for experimenting with New Salem—in effect, that experiment had already been conducted in the decades before the hunting began. And though regeneration was robust in New Salem, the levels of browse showing up in field surveys suggested that there were still plenty of deer in the area. Moreover, fewer than half the deer killed in New Salem during the 1996 and 1997 seasons were female, as opposed to the average of 60 percent females shot on the reservation over the seven years of the hunt. From the beginning, the hunt had been designed to reduce the number of females, the key to reducing and then stabilizing the size of the herd as a whole. But the sex ratio of the New Salem herd continued to favor

---

*The situation, already overbrimming with irony, thus produced another. The Division of Fisheries and Wildlife, the bête noir of the opponents of the hunt who were certain that with hunting and fishing now allowed the division basically would bully the MDC into turning the reservation into a sportsman's paradise, was in complete accord with the MDC's desire to continue hunting in all five areas of the reservation. Increasingly, the division is compelled to respond to nuisance animal reports, many involving white-tailed deer. Whatever their historic bias toward providing "targets" for the sportsmen and women whose license fees fund their work, their mission and self-definition is changing from a "give-em-targets" mentality to a much more sophisticated ideal of stewardship that has gone largely unheralded—for reasons we shall explore later.

population growth. To make matters worse, the winter of 1997–98 was unusually mild and there was a large mast crop. This almost certainly meant that the does entered the winter in good condition and gave birth to healthy fawns in the spring of 1998. With hunting suspended for the fall, and with plenty of browse available from the regeneration that has been occurring, the herd in New Salem is poised for robust growth. Thus, the outcome of this "experiment" is likely to be far more satisfying to hunters than to the foes of hunting who pressed for closing New Salem. The ideological commitment of Massachusetts Audubon and the MSPCA to an unmanaged nature will in all likelihood be trumped by deer biology, and this will mean the eventual resumption of hunting in New Salem.

The hunters who sided with the opponents of hunting should take little comfort from the prospect of one-upping their foes. Though they may "win" in the short run, in the longer run their shortsightedness may be every bit as self-defeating as the ideological rigidity of the MSPCA and Massachusetts Audubon. If large numbers of hunters begin to insist that the deer herd on the reservation be managed to maximize opportunities to shoot deer, they will do themselves a serious disservice and may well wear out their welcome thoroughly. The public's patience with hunters whose exclusive interest is in having animals to shoot at is wearing thin. The Quabbin hunt presents hunters with an opportunity to demonstrate that they not only can be safe and considerate but also informed and engaged environmentalists willing to enlist in the management of a precious natural resource even though this may mean fewer opportunities to kill a deer.

Some hunters seem to have gotten this message, judging from the numbers of hunters who have applied for the hunt year after year, even though the annual numbers of deer killed were tailing off sharply. To be sure, the number of hunters applying for permits to hunt the Quabbin declined steadily after the first couple of years. In the first year, over seven thousand hunters applied for a little over one thousand permits. Applications rose in 1992, no doubt reflecting keen interest in the Prescott Peninsula, long regarded as one of the "wildest" and most scenic of the areas of the Quabbin and one to which public access had long been most restricted. Just over two thousand of these applicants were chosen. Ever since, the number of applicants has been declining.

In 1997, 1790 hunters applied and 1525 were selected. It is unclear what, if any, effect the withdrawal of New Salem from the 1998 hunt will have on applications, but it is unlikely to boost the number.

The MDC is a bit worried about this trend, but there are some virtues concealed in the shrinking applicant pool. It is likely that as the challenge of the hunt has risen, with the deer no longer either super-abundant or virtually tame, the motives of the hunters who apply are different. The novelty of hunting in a seemingly pristine area has worn off, and the likelihood of coming home with a deer is now more nearly the same as hunting anywhere else in Massachusetts. Thus it is proba-ble that many of the people who now are applying to hunt the Quabbin appreciate the virtues of a controlled hunt—assignment to a particular area which reduces competition as well as the risk of accident, and not having to worry about gaining several different landowners' permis-sion—as well as the satisfaction of knowing that they are assisting in a worthy endeavor.

Still, if the number of applicants continues to decline, it may be necessary to adopt a system that permits both individuals and groups to participate in the hunt. It might also become necessary to cultivate actively a core group of hunters who can be depended upon, year in and year out, to concentrate their hunting efforts on the Quabbin, becom-ing, in effect, more nearly full-fledged members of a management team.

In short, though the hunt is having its intended effects and the an-nual hunts have gone smoothly, if not always flawlessly, there are plenty of sources of concern. Opposition to the hunt and, more broadly, to the management of nature, remains institutionalized and persistent, de-spite repeated confirmations of the soundness of the Quabbin man-agers' efforts.* The problem, of course, is that the success of the Quab-bin managers does not lead to the stability and self-management for

---

*While Mass Audubon and the MSPCA were trying to subvert the Quabbin management plan, the Forest Stewardship Council, which calls itself an "international, non-governmental organiza-tion which promotes the certification of forest management which is environmentally sound, socially beneficial, and economically viable," was drafting strict standards by which forestry operations could be certified, with the wood products labeled so that consumers can be assured that they are purchasing lumber taken from well-managed forests. Another independent group, SmartWood, evaluated the Quabbin forestry operation using the FSC draft standards and gave the Quabbin its highest rating, sort of the equivalent of the Good Housekeeping seal of approval.

which the critics yearn. A healthy and diverse forest will always be fine deer habitat. With moose added to the mix, the challenges the MDC faces increase. It is hard even to contemplate the hue and cry that would go up should the managers some day conclude that moose, too, will have to be hunted. The mere mention of a moose hunt caused eyes to roll among the Quabbin personnel.

There are other troublesome matters that make the management of the Quabbin complicated. Beaver and muskrat populations appear to be on the rise, as they are throughout much of the rest of the state. When I talked about beaver with MDC staffers in 1990–91, the beaver were at a low ebb, the deer having effectively deprived them of habitat. With fewer deer and more regeneration, they knew the beaver would eventually stage a comeback. But this was not a huge concern then because the MDC did not have to seek authority to trap beaver and so controlling the beaver population could be done well below the radar screen of the public. That was 1991. In 1996, the MSPCA, the Humane Society of the United States (HSUS), Massachusetts Audubon, CEASE, and a variety of other animal rights groups carried out a very effective ballot initiative campaign aimed at eliminating trapping with leghold or body gripping traps. The measure passed by a margin of two to one. Now that the only traps legal for use in Massachusetts are the Bailey and Hancock traps, controlling beaver and muskrat is a far less simple matter. Both are large, cumbersome affairs; worse than the inconvenience in using them is their lowered effectiveness. When the beaver become numerous enough to be a serious problem, the MDC will almost certainly seek exemption from the ban on leghold traps in a process that requires public hearings. This will embroil the agency in yet another highly charged debate.*

The ban on traps, like the initial controversy over the deer hunt and the forestry practices of the MDC, reflects a dramatic shift in how the

---

*It may well be the case that work on this ballot initiative drew the energies and attention of animal rights activists away from the Quabbin deer hunt. In any event, the activists will be aroused when, as seem likely, the MDC will need to control beaver more aggressively. The irony is that regardless of the method of trapping, whether with live traps or body gripping traps, beaver will be killed. State law forbids the transport and release of wild animals, so the "humanely" trapped animals will be destroyed.

public views nature and natural resources. A generation ago, those who thought about the natural environment generally worried that we were not paying sufficient attention to the warnings of environmental scientists and resource managers. Bent on raising standards of living, we were not conserving precious resources. Attention mostly centered on non-renewable resources like oil, but there was also alarm sounded about exceeding the capacities of our forests, oceans, and rangelands to keep up with our intensifying harvests. Though our wasteful ways have continued, largely unabated, more and more Americans have begun to question these practices. When I began writing *Going Wild*, this shift was already well under way. It has gathered even more momentum since, with the result that episodes like the controversy over the Quabbin deer hunt are fast becoming commonplace. Though I did not fully appreciate it at the time, the Quabbin controversy is a chapter in a much larger story. It is to this larger picture that I would like now to turn.

Loss and Hope

For much of the past 150 years, roughly the period in which the genre of nature writing developed and during which modern environmentalism took shape, Americans who cared to listen have heard a mounting litany of woe. George Perkins Marsh set the tone when he observed that our ignorance and indifference were upsetting the delicate balance that gives the physical world its integrity. In our wake, Marsh saw a turbulent mix of disruption and ruin. On his world travels as well as from his farm in central Vermont, he watched thickly forested hillsides denuded, replaced by farms and grazing animals. The trees which once shielded the land were turned into lumber and fuel, the latter filling the air with acrid smoke. The spread of agriculture prevented reforestation, depriving wildlife of vital habitat at the same time that wildlife populations were being heavily exploited for meat, fur, hides, and feathers. The physical world around him was threatening to become, like Humpty Dumpty, shattered into a jumble of fragments utterly lacking in coherence. Ever since, nature writing has been framed by this narrative of loss. Even when the writing was rhapsodic, as was Muir's, the spectacles and wonders evoked could not help impressing upon the reader the widening circle of disruption and degradation against which Muir's beloved Sierras seemed even more compelling. This sense of loss lent urgency, in Muir's day as in our own, to the desire to protect those few places yet to bear the full brunt of civilization.

To be sure, there was a counterpoint to writing about nature in elegiac mode. John Burroughs, a contemporary of John Muir and the most widely read nature writer in the decades around the turn of the century, wrote compellingly about the marvels revealed by a close observation of nature, including even the highly modified nature of one's own back yard or nearby vacant lot. Though Burroughs traveled widely, he did not insist, as Muir did, upon making invidious comparisons between nature in farm country and the pristine high Sierras. Burroughs insisted that nature is everywhere and on all scales irrepressibly bringing forth life. Our back yards cannot compare with the spectacle of Yosemite Falls, but the delicacy of a moth's wing is every bit as spellbinding at home as it is in some sublime setting. Burroughs was not an apologist for exploiters of nature any more than he was indifferent to the damage industrialism and urbanism were inflicting upon the landscape. Indeed, his voice was one of the more persuasive in the mounting chorus demanding that our natural resources be protected from heedless overexploitation. He was an ardent supporter of the Adirondack Park, not least because it embraced a mix of preserve and prudent use. He aligned himself with Theodore Roosevelt's and Gifford Pinchot's efforts to institutionalize the ethos of stewardship in our nation's utilization of natural resources.

Echoes of Burroughs's inclination to temper loss and degradation with accounts of nature's resilience continue down to the present, even though they are often framed within the narrative of loss. Scott Russell Sanders and Robert Sullivan each have engagingly insinuated an ironic note into the dominant narrative of loss.[2] Sanders recalls growing up on an army base in Ohio where, along with spent ordnance, decommissioned jeeps, and sundry leavings of army life, all sorts of wildlife abounded, seemingly indifferent to the target practice and war games going on around them. Sullivan's chosen landscape is an even more unlikely place in which to appreciate nature: the Meadowlands, just across the Hudson from New York City, is one of the most alternately abused and neglected patches of territory in the New World, right up there with Butte, Montana, and Sudbury, Ontario. Even a talent as large as Burroughs would have had trouble finding anything inspiring, much less redemptive, in the Meadowlands. And yet, with considerable irony, Sullivan finds nature at work, in places appearing to have begun to neutralize if not erase the most obvious evidence of our abuse. Like salmon defying currents to get upstream, microbes team up with flora and fauna to reclaim what they can for their own. There are impediments to this reclamation, of course. However generative and resilient nature might be, we have put some impressive barriers in its way, just as we have erected dams for which the salmon is no match. The Meadowlands will never be what it was before we began dumping, filling, and otherwise defiling the area. But if we can keep from insulting it further, if we could even give nature a bit of a helping hand, the Meadowlands might not forever remain a monument to our greed and shortsightedness.

Army bases, dumps, and fields laced with chemicals are irrevocably altered, these writers tell us, but there is nevertheless a constant generative force at the heart of nature that is as undaunted by our sea walls and dikes as it is by our heedless profligacy. And many creatures, especially small ones, actually can thrive on our modifications of the environment. The late Vincent Dethier, a zoologist, demonstrates this charmingly in his book, *The Ecology of a Summer House*. Dethier catalogues the creatures, from wasps to mice and bats, who share his cottage on an island off the coast of Maine and seem not the least inconvenienced by human intrusion. Of course, there have been real and permanent losses, but that is not the whole story. Indeed, even so confirmed an elegist as Bill McKibben admits some small ray of hope in his recent collection of essays, *Hope, Human and Wild*.

These two narratives, one of loss, the other of recovery, have dominated our thinking about nature and have shaped our contradictory and often embattled attempts to define an ethically and materially sustainable relationship with the natural world. In their extreme versions, these two narratives are diametrically opposed. The narrative of loss becomes a wholesale condemnation of modern society and an evocation of a fast-approaching apocalypse. At the other extreme, the narrative of recovery

can provide a fig leaf of respectability for the so-called Wise Use movement and others who reject virtually all environmental regulation and
restraint on exploitation. The narrative of recovery also underpins those
few scholars who remain convinced that nature's bounty is, for all
practical purposes, unlimited, even capable of absorbing very large
increases in human population.[3] Between these poles, however, there is
considerable flux between optimism and pessimism and lines are easily
blurred. And yet, for all the mood swings, one of the truly remarkable
features of an otherwise polarized and volatile public has been the
consistency of the American public's concern over the environment.
The narrative of loss has gotten peoples' attention and made them
worried.

For the past twenty-five years, public opinion polls have recorded
strong support for spending tax dollars on the environment. For example, the National Opinion Research Center (NORC) at the University
of Chicago has been asking carefully drawn random samples of American adults about their support for environmental spending nearly every
year since 1973. In only two of twenty-five years between then and now
has support for increased spending on the environment dipped close to
50 percent. In most of the past twenty-five years, two-thirds to three-
quarters of Americans said that the United States was spending too
little "on improving and protecting the environment." This willingness
to commit more tax dollars to the environment is all the more impressive given the prevailing anti-tax, anti–government-spending mood of
the country. In fact, public support for environmental spending rose
sharply through the Reagan-Bush and Bush-Quayle era (see chart 1 in
the appendix).

It also should be noted that support for spending on the environment is evenly spread throughout the population: there is no gender
gap and differences along racial lines are small compared to black-
white differences on most other policy issues. Only when we look at
age do we find appreciable variations in levels of support for increased
spending on the environment. As one might expect, young Americans
are far more likely to think we are spending too little on the environment than are older Americans. Seventy-four percent of those under
thirty say we are spending too little while only 47 percent of those fifty
and older think too little is being spent.

In 1993, 1994, and 1996, NORC asked a range of more detailed questions about Americans' attitudes toward the environment that go beyond a general endorsement of higher spending. Over half of the respondents in 1993 and 1994 claimed that "I do what is right for the environment, even when it costs more money or takes up more time." Once again, the only variation of note in this response comes with age, but here things are reversed: surprisingly, the younger respondents are less likely (46 percent) to say they "do the right thing" than their older counterparts, 63 percent of whom say they try to do what's right.

To be sure, all is not sweetness and light. When the questions get more pointed, environmentalism starts heading south. Barely 50 percent of those polled in 1993–94 said that they would be willing to pay "much higher prices," a little more than a third said they would be willing to pay "much higher taxes," and less than one-third said that they would accept cuts in their standard of living "in order to protect the environment" (see chart 2 in the appendix).

It is tempting to interpret this cynically and conclude that since Americans are not prepared to put their money where their mouths are, there is little reason to pay attention to their pious utterances. This would be a mistake. However important money may be, it is not the only measure of commitment to protecting the environment. For example, when they were asked in 1996 whether or not they agreed with this statement—"Natural environments that support scarce or endangered species should be left alone, no matter how great the economic benefits to your community from developing them commercially might be"—60 percent of the respondents agreed. Unfortunately, we do not have a time series on this question, but it is hard to imagine a representative cross section of Americans giving assent to this question in, say, 1950. Indeed, it is doubtful that anyone would have dreamt of asking such a question in 1950. Even more impressive, especially with all the gnashing of teeth over government intruding on the rights, especially the property rights, of Americans that has characterized public discourse for the past twenty years, Americans are surprisingly willing to accept regulations aimed at protecting the environment. For example, 89 percent endorsed the following statement: "It should be the government's responsibility to impose strict laws to make industry do less damage to the environment." More astonishingly, 73 percent agreed that "for certain problems, like environmental pollution, international bodies should have the right to enforce solutions." It would appear that only a handful of Americans are worried about the United Nations flying black helicopters over Idaho and Montana.

If this were all there was to the story, the matters before us would be simple: Americans, by a large margin, endorse solicitude for the environment, even though their resolve wavers when that solicitude has some bite—when it takes something out of their pocket or puts a crimp into their life style. The plot thickens, however, when we consider what people have in mind when they embrace nature. The shift in the way Americans view nature goes well beyond wanting more trees and fewer strip malls. Appetite for the *wild* has been awakened, almost as if large numbers of our fellows had read and taken to heart Thoreau's memorable assertion, "In wildness is the preservation of the World."

Thoreau was not speaking for his contemporaries when he praised wildness. They were still busy pushing wildness as far away as possible, preferring instead clear boundaries, well-kept fields, and domestic animals safe from wildlife that threatened or competed with them. By the beginning of the twentieth century, perhaps because the prospect of wildness holding back the march of civilization no longer seemed nearly as threatening, steadily growing numbers of Americans sought out remote areas which, even though not wild in the fullest sense, stood in sharp contrast to the cities and suburbs from which the sojourners hailed. Hiking and camping, as well as fishing and hunting for recreation, as opposed to subsistence, became very popular, and once paid vacations became widespread, visits to state and national parks and forests rose sharply.

Thoreau's wild was an abstraction, a metaphor really.* Ever since Thoreau, people drawn to the wild have imagined that by immersing themselves in nature, they come face to face with eternal truths, not least of which is an understanding of how the natural world is intricately stitched together in elaborate webs of symbiotic relationships.

---

*In fact, Thoreau's experience in the deep woods of Maine did not exactly inspire him. He was uncomfortable with the remoteness and even more with the impenetrable tangles of trees, vines, and blowdowns that impeded his movement as well as his capacity to see much beyond the end of his nose.

The wild has come to represent all that is pure and innocent, in sharp distinction to the earlier view that the wild was Satan's sanctuary. In the early stages of this shift in perspective, the desire to preserve at least portions of what remained of the wild got folded in, albeit with some creases, with the emerging ethos of stewardship. The idea was to conduct ourselves in ways that would sustain natural diversity and yield a continuing harvest of resources, both material and aesthetic—to preserve here, prudently use there, and, where possible, promote the idea of multiple use so as to broaden the ranks of those who have a commitment to the wild.

In recent decades, the reverence for the wild has increased and stewardship has become suspect, at least in part because stewardship has frequently seemed to be joined at the hip with the forces intent on boosting consumption and intensifying the harvest. In effect, the narrative of loss has turned the dominant national celebration of Manifest Destiny and growth on its head—the march of progress is now commonly depicted as heading us for a cliff and the state and federal agencies managing our natural resources are characterized as villains, if not leading the march at least playing an active supportive role.

Again, data from NORC's General Social Survey, 1993 and 1994 are instructive. A solid majority of Americans in each of the two annual surveys agreed with the statement, "Almost everything we do in modern life harms the environment." More important, Americans appear to have lost confidence in our capacity to solve environmental problems by relying on science and technical know-how. Only 20 percent of the respondents in both surveys thought that "Modern science will solve our environmental problems with little change to our way of life." No doubt this reflects what seems to be a growing skepticism about science. Fifty-five percent of all respondents agreed with the statement, "We believe too often in science, and not enough in feelings and faith." And, as if to emphasize the importance of faith, nearly 80 percent said that "Human beings should respect nature because it was created by God." Given this set of beliefs, it should not be surprising to learn that just over half of all respondents reported believing that "Nature would be at peace and harmony if only human beings would leave it alone" see chart 3 in the appendix.

Data collected recently by a team of anthropologists allow us to fill in more details about the ways Americans think about the environment.[4] They assayed the "environmental values" of five groups of people: members of EarthFirst!, members of the Sierra Club, the general public, workers in the dry cleaning industry and, finally, workers in sawmills.* Though, as one would expect, there are some dramatic differences between the five groups of respondents, what is even more impressive is the degree to which there is consensus on precisely the matters we have been discussing (see chart 4 in the appendix for the complete report). For example, almost everyone agreed that "We have a moral duty to leave the earth in as good or better shape than we found it." Sawmill workers were the only group in which fewer than two-thirds agreed that "Nature is inherently beautiful. When we see ugliness in the environment, it's caused by humans." And they were well outside the fold.

Ranks closed again on the following statement: "Nature may be resilient, but it can only absorb so much damage." Ninety four percent of EarthFirst!ers agreed, as did 85 percent of sawmill workers. Similar accord was found on this statement: "Nature has complex interdependencies. Any human meddling will cause a chain reaction with unanticipated effects." 97 percent of EarthFirst! members agreed, as did nearly two-thirds of sawmill workers (63 percent). Sawmill workers split from the pack on the question of extinction. Large majorities of the other four groups, ranging from 78 percent to 97 percent, agreed that "Preventing species extinction should be our highest environmental priority. Once an animal or plant species becomes extinct, it is gone forever." Only 41 percent of the sawmill workers agreed. Sawmill workers edged back into the fold, though they still lagged well behind the others when the issue of extinction was rephrased: "All species have a right to evolve without human interference. If extinction is going to happen, it should happen naturally, not through human actions."

Large majorities of all five groups also agreed with the following two

---

*The samples from each group were not random and so generalizing beyond the respondents themselves is not warranted. Still, the results Kempton, et al. report are certainly consistent with the findings derived from carefully drawn random samples such as those of NORC.

statements: "Humans are ripping up nature, feeling that they can do a better job of managing the earth than the natural system can"; "Humans should recognize they are part of nature and shouldn't try to control or manipulate it." Given this, it should come as no surprise that hardly anyone in any of the five groups had much faith in technological fixes. The most optimistic were, of course, sawmill workers, but even they could only muster 15 percent who agreed that "We shouldn't be too worried about environmental damage. Technology is developing so fast that, in the future, people will be able to repair most of the environmental damage that has been done.

In this context, it is easy to see why the public is increasingly turning against those who log or who argue for the active management of wildlife, especially when management means killing overabundant animals either because they are damaging the environment or are interfering with or endangering humans. We have gone, in effect, from casual disregard of nature to worshipful awe.

Americans have long thought of nature as a storehouse or a cupboard from which they could draw all that suited them. Beginning around the middle of the nineteenth century, after a long orgy of ransacking the cupboard, Americans began debating whether access to the cupboard ought to be restricted and regulated and how this could be accomplished. The conservation movement, as it came to be called, tried to encourage restraint: making fewer and more careful trips to the cupboard and resisting the impulse simply to empty the shelves. The idea, as put forward by men like Gifford Pinchot, was not to shut the cupboard door but rather to end wastefulness and uncontrolled exploitation of natural resources. With better scientific understanding and improved techniques of resource management, Pinchot and his followers were confident that consumption and replenishment could be balanced. The rhetoric of conservation and of sustainable yields reassured Americans that the cupboard was safe. In the last three decades, however, support for this venerable notion of conservation has waned, gradually at first, but then accelerating rapidly. In its place, the desire to treat nature less as a cupboard and more as a "living museum" has grown. Not too long ago environmentalists thought of themselves as stewards who watched over nature lest its regenerative capacities be compromised. Now environmentalists are just as likely to think of themselves as curators: like all curators, they want to add to their "collection" or, at the very least, keep what they have intact.

The continuing conflict over the management of our flagship park, Yellowstone, though more complex, nonetheless parallels the Quabbin controversy. Under the leadership of Starker Leopold, a son of Aldo Leopold, the National Park Service adopted a system of "natural regulation" for Yellowstone which has been in force since the late 1960s. The aim was to allow nature free reign on the assumption that, if left alone, the park would in time come to resemble the landscape that greeted the first white men who looked out over the Lamar Valley. Indeed, park historians and biologists worked hard to figure out just what this "original condition" might have been. It turns out to be no simple matter, not least because a lot of crucial evidence is missing. As with so many things about the natural environment, we do not have good data over long enough stretches of time to know if this or that condition is typical or transient, part of a long repetitive cycle or a chance event.[5]

This uncertainty means, in the final analysis, that favoring natural regulation is really a gentle way of saying that we should be prepared to take what comes, a point many of the critics of the MDC made: instead of imposing our preferences on nature, we should learn to accommodate to whatever nature sends our way. Natural regulation depends upon our willingness to accept fires, floods, periodic heavy winter kills, blights of one kind or another—the usual suspects. In general, we have been willing to "let things go" only in places that don't otherwise hold much interest, places like vacant lots, degraded landscapes, or acreage that offers no commercial or recreational attractions. But in places like the Quabbin, where a water supply is in the equation, or Yellowstone, a destination for millions of summer vacationers intent on seeing some of nature's most wondrous spectacles, a principled indifference to the erratic whims of nature is hard to sustain.

Proponents of natural regulation see silver linings where others are inclined to see dark clouds. The fires that swept across Yellowstone in 1988 are now heralded as unleashing a virtual flood of regeneration and species diversification. The staff at Yellowstone has labored long and hard to assuage the dismay of park visitors when, even ten years after fires swept the park, they look out on a scene whose most striking feature, from a distance, is charred remnants of once majestic stands of lodge-pole pine. Beneath these skeletal remains is a verdant carpet of early successional growth. Although that's not what tourists notice, those who are knowledgeable do see the wildflowers, brambles, and young shoots as harbingers of recovery. John Varley, the longtime chief of scientific research at Yellowstone, is struck by what he sees growing now that the dense canopy has been opened to the light. "I was hiking the east side of the park," he reports, and "I ran into mountain holly-hock. I've been here twenty years, and I've never seen a single one. There was a hillside covered with their lavender blooms—so many of them it was too thick to walk through."[6] And there is surely more food for the large mammals who inhabit the park.

It is instructive, in this context, to reflect on the difference between the way in which fire and logging have been treated by the environmental press. Clear-cuts are excoriated—they are routinely depicted as a scourge, creating a wasteland where once there was abundant life. Not only are clear-cuts called an abomination to the eye, they are also held up as examples of how wantonly we are stressing the regenerative capacity of the earth.[7] By contrast, fire is now celebrated as nature's way of renewal, at least when the fire is sparked by lightning, not careless campers. I would not claim that fire and clear-cutting are equivalent, but in most instances I would wager that the different biological impacts of cutting versus fire are far smaller than the blanket condemnation of the former and the blanket embrace of the latter would suggest.*

The visual results are certainly similar. But imagine the outcry had it been chainsaws, not lightning, that had made way for the explosion of biotic variety and vitality that Varley and others are thrilled to be witnessing. The lodge-poles pine will return, in time. Experts estimate that roughly two hundred years from now, maybe even sooner, visitors to Yellowstone will once more gaze out over an expanse of tall pines. In

---

* I fully appreciate the claim that fire and clear-cutting have different ecological implications—and that, to make matters more complicated, the implications are different in different regions and climate zones and for different soils and topographies. Still, it is worth reflecting on our recent willingness to accept fire and reject logging.

New England, it takes half that time for an area to become reforested. (Remember, the Quabbin was heavily cut over—and then blown over by the hurricane of '38—a mere sixty years ago, and it is now admired for its wilderness qualities.) And yet, opposition to *clear-cutting* is as intense in New England as it is in the West. Storms, fire, disease, and saws are all "enemies" of a forest and each introduces different stresses and produces a different kind of vacuum. But, ironically, were it not for these sources of disturbance, our forests would be host to a much narrower range of flora and fauna than we now expect to find.

Yellowstone, because it is a park, can be left to natural regulation if the public is willing to take what comes. It is certain that whatever comes, there will be plenty to stimulate the senses and plenty of things that will engender awe in the presence of nature's irrepressible drive to fill in empty spots. This is what some parks, at least, should be all about. But as a general rule, letting nature decide will not work, not even in most parks. We must cut trees; we must keep land cleared for agriculture; we must manage wildlife, which sometimes means protecting them and their habitats and sometimes means killing some of their number. Nature doesn't care if the deer denude a hillside, but we surely ought to.

This said, there is little doubt that we should expand the zone of our indifference to what nature sends our way. For example, we should resist the temptation to build in flood plains or along fragile coastlines or in areas that are otherwise unstable. At the very least, we should not encourage such building by offering federally underwritten insurance against the perils of flood and beach erosion. We should not expect nature everywhere to be made to conform to our wishes. Just as surely, though, we should not be lulled into thinking that only untouched nature is worthy of our appreciation.

If the proponents of natural regulation had only such modest goals as these, only the most enthusiastic developers would object. But, like the critics of the MDC, proponents of natural regulation go further: they wish to see landscapes purified of human presence. Humans are, in effect, written out of the equation. Before natural regulation became official policy at Yellowstone, elk were killed in the park in order to keep their numbers in check. In the late 1960s, when the herd was estimated to be less than five thousand animals, management of the elk herd was ended, at which point the herd grew rapidly, reaching as many as nineteen thousand by 1988. Then came the fire, followed by a severe winter, and the elk, deprived of browse and shelter by the fire, suffered a huge winter kill. But it was not long before regeneration began providing browse again and the herd quickly rebounded, and now numbers, according to Paul Schullery, a noted historian of the park, at least twenty thousand. Though there is still intense debate about whether the park can sustain this many elk without their having adverse effects on other aspects of the park's ecology, there is little disagreement over the fact that the size of the elk herd reflects a key missing ingredient in the Park's ecosystem: predation.

Wolves, along with humans, historically exerted a major influence on the elk. Wolves began to be hunted, trapped, and poisoned all across the West in the early decades of this century in a systematic campaign to rid the range of this predator in order to make the world safe for livestock. Though early conservationists, including the elite hunting organization Boone and Crockett, opposed the effort, anti-wolf hysteria overpowered reason and wolves were pursued mercilessly, even when livestock were not at issue, as in Yellowstone. By the late 1930s, the wolf was eliminated from the ecology of Yellowstone, and with much tighter control of poaching as well, only the annual harvest of elk by park personnel kept the herd in check. When that practice ended, and there was nothing to stop the elk from increasing their numbers, the idea of bringing wolves back to the Park soon began to surface.*

The saga of restoring wolves to Yellowstone is a compelling one, at times more a courtroom drama than an ecological project. There are many fine accounts, among the most readable of them Thomas McNamee's *The Return of the Wolf To Yellowstone.*[8] We need not consider the full story here, but we should note one crucial feature of the

---

*Although supporters of the wolf restoration think that wolves will control the elk herd, no wildlife biologist whose work I have read or with whom I have spoken is sure that such a plan will be successful. Indeed, the experience in Minnesota with wolves and white-tailed deer clearly indicates that wolf numbers and deer numbers are, at best, very loosely correlated. The relationship between predator and prey is, as Botkin and others have noted, by no means straightforward. But this complexity has not kept partisans of wolf reintroduction from intimating that reintroducing wolves will return things to balance.

thinking behind the idea of natural regulation. McNamee, a rancher, former president of the Greater Yellowstone Coalition, and a strong defender of the wolf reintroduction, describes the park and its environs as follows: "Yellowstone Park is . . . the geographic, ecological, and spiritual center of the largest remaining essentially intact ecosystem in the temperate zones of the earth—the heart of an 18-million-acre complex of wildlands. . . . It is recognized here that the single most powerful absence for the Greater Yellowstone Ecosystem—what demands the cautionary adverb in 'essentially intact'—is the absence of the ecosystem's only missing component, the wolf." The wolf is so powerful a symbol that it is easy to overlook the fact that McNamee, no doubt without malice, conveniently writes humans out of the definition of "intact ecosystem." In his ecological zeal, he has done what the earliest European settlers did when they described this continent as a vast, *empty* new world.

Of course, the New World was no such thing. Humans had been exploring and exploiting virtually every nook and cranny of the continent long before Europeans even dreamed of the New World. And this is precisely the rub—to speak of intact ecosystems without factoring in the ways Homo sapiens have, for the past ten thousand years or so, intentionally and unintentionally altered things is both bad history and bad ecology. In our present mood, this is precisely what happens over and over again—things can be right, it is assumed, only if humans are erased from the scene.* Once we get all the pieces put into an order that suits us, then we can sit back and admire the living museum, conveniently overlooking the fact that we are admiring our handiwork as much as we are witnessing nature's wonders. Moreover, preserving habitat with the presumption of intactness leads people to believe that once stewardship is set aside, the need for it disappears. After all, intact ecosystems take care of themselves, don't they? Consider the wolf yet again.

---

*Lest there be any misunderstanding, I am not attacking wolf reintroduction. Far from it. I think the extirpation of the wolf was a sad and foolish thing and I support efforts to reestablish wolf populations. But as will be clear shortly, reintroducing wolves will not relieve us of the need to manage wildlife. In fact, it will simply add wolves to the already growing list of wildlife species requiring management.

The successful reintroduction of wolves to Yellowstone Park will likely result in a more biologically dynamic situation as well as greater biological diversity.* But it will also result in the steady dispersal of wolves beyond park boundaries and, eventually, into ever more marginal habitat and into closer proximity to cities and towns. The reason is simple: wolves are territorial and hierarchical. This means that adolescents are sent packing, even if there is more than enough food within the parents' territory to support an ever larger brood. Abundant food, such as the park will provide for a long time, reduces the territorial needs of each pack, so the park for some time is likely to be home to a greater density of wolves than in northeastern Minnesota, where the largest natural population of wolves in the lower forty-eight resides. In Minnesota, packs require between fifty to one hundred square miles, roughly, but defended territories have been known to be as large as two hundred square miles. Pack size also varies as a function of food supply as well as variations arising from the social dynamics of particular packs and idiosyncracies of pack members. Most packs do not exceed eight individuals (though packs as large as twenty-one have been recorded).

The fact that there is great habitat for wolves in Yellowstone has meant that the reproductive rates of the transplanted wolves have been high and is likely to remain high for years to come.† Whatever variation develops from year to year will be a function of weather, disease,

---

*Although success has to be tempered a bit. Even though the biological fate of the wolves seems quite secure, their legal fate is clouded. In a suit brought by cattle interests and partisans of Wise Use, to which some environmental groups added their support, a federal judge found in favor of the opponents of the wolf reintroduction. Then the U.S. Fish and Wildlife Service, in order to gain support from ranchers, agreed to call the transplanted wolves an "experimental population." This meant that the wolves were not protected by the provisions of the Endangered Species Act and, should a wolf be found killing livestock, it could be killed. Were it protected by the ESA, killing it would be a serious crime, no matter what it had done. The plaintiffs argued on appeal that there were some indigenous wolves on the scene and that it was impossible to distinguish between the "natives" and the "transplants." Lowering the protection of the transplants necessarily lowered the protection of the natives, who truly were endangered. Thus, they claimed, the reintroduction violated the ESA. As of this writing, the case is still on appeal.

†Of course there will be year-to-year variations in wolf numbers, just as rates of reproduction will vary. An especially hard winter or late spring can affect litter size and survival of newborns. Territorial disputes between packs will also affect wolf numbers. One pack has already been decimated by another pack. But as long as elk remain abundant, the wolves in the park will prosper.

fluctuations in the abundance of species that wolves prey upon, and wolf mortality from other sources, such as other wolves or competing predators. (Nineteen ninety-eight, for example, appears to have been a poor year for wolf reproduction.) High rates of reproduction will inevitably mean that young wolves will be on the move seeking territories of their own. Though there can be no certainty about when wolves will begin dispersing beyond the park (as opposed to sojourning outside the park, which has already occurred a number of times), or in what annual numbers they will be leaving, there is no doubt that dispersal will occur. How will these wolves fare, far from their "intact ecosystem" home in Yellowstone? Will the broad public support currently enjoyed by the wolves continue when wolves begin showing up in the suburbs of Denver? Whatever their fate, they will almost certainly be the center of disputes not unlike the struggle over deer at the Quabbin.

The wolves will dramatically—but hopefully not tragically—illustrate a point all too easily overlooked: there are no borders around so-called intact ecosystems, a fact which calls into question the usefulness of the term ecosystem itself. Moreover, if ecosystems are not bounded in any rigorous sense, then concepts such as intactness or integrity begin to lose their bite. Our fascination for the seemingly intact wild obscures the larger challenge we face: the challenge of figuring out how to be better stewards. Sooner or later, the Yellowstone wolves will pose exactly the same predicament that wildlife officials in Minnesota face.

When the Endangered Species Act was passed and the Minnesota wolves were listed as endangered, there were several hundred wolves in Minnesota, concentrated in the deep woods north and west of Lake Superior along the Canadian border. Now, twenty five years later, there are several thousand wolves in Minnesota and several hundred more in northern Wisconsin and in the Upper Peninsula of Michigan, these latter populations founded by wolves dispersing from northern Minnesota. (One wolf is known to have dispersed 550 miles away from its home base in northern Minnesota!) The wolves have expanded in successive concentric rings south and west from the northeastern "seed bed" and now are within an easy day's lope of Minneapolis-St. Paul.

David Mech, the best known of the nation's wolf experts and a close student of the Minnesota wolves, shocked a recent gathering of experts and citizens interested in the prospect of reintroducing wolves to New York's Adirondack Park when he bluntly noted that three hundred to five hundred wolves would have to be killed annually in Minnesota if wolves were to be kept safely away from the Twin Cities and from the farms that blanket the southern part of the state.[9] In short, having protected the wolves in habitat that was ideal for them, we are now faced with agonizing and almost certainly deeply polarizing management choices. Preserving wildlife and their habitat does not get us out of the business of stewardship, but the preoccupation with preservation clearly has made stewardship more controversial, more politicized, and thus much more difficult. To make matters worse, the successes we have had in bringing back wildlife species and in restoring habitat have made stewardship all the more necessary, precisely as the stewards get more embattled.

A similar point was made recently by William Cronon, one of the nation's leading environmental historians. In "The Trouble With Wilderness; or, Getting Back to the Wrong Nature," Cronon questioned the wisdom of the environmental movement's long preoccupation with wilderness, particularly the tendency to define worthy landscapes almost entirely by their closeness to the nineteenth-century ideal of the sublime: craggy peaks overlooking a serene, almost pastoral, verdant valley, a heady mix of the overpowering force and benign fecundity of nature. With our eyes thus filled, Cronon averred, we are less likely to be moved to value the more prosaic landscapes in which most of us live and work. Compared to the Grand Tetons or the Lamar Valley or Yosemite, most places look hopelessly uninspiring if not defiled.

For his pains, Cronon has been roundly excoriated in the environmental press. What we need, his critics all seem to agree, is more wilderness. Among the most thoughtful of Cronon's critics is a colleague of his in the environmental studies program at the University of Wisconsin, Madison, botanist Donald Waller. Waller makes a strong case for creating very large, and wherever possible *unbroken*, expanses of "wildlands," land that is exposed to the bare minimum of human presence and, if it is to be managed at all, managed in the most minimal way possible.[10] Waller's wildlands, while not exactly "wilderness," share important features with wilderness: the presumption of intactness, the removal or drastic reduction of human presence, and reliance on the capacity of nature to be self-organizing and naturally regulated.

On the face of it, Waller's proposal seems compelling—who would not want to see large swaths of prairie stretch once more across the Midwest or large unbroken tracts of mature forest such as greeted settlers' eyes as they spread west into Michigan, Wisconsin, and Minnesota? And who among us does not share Waller's curiosity about what happens in such places over time? How do plant dynamics differ from conventionally disturbed areas? What happens to deer populations or song birds? Do undisturbed areas generate more or less biological diversity than areas less insulated from human activity? As laboratory, inspiration, or balm, wildlands have much to offer. But then come the mundane, practical issues. Where do we find the money to buy the land or compensate owners for restricting the range of uses to which they can put their land? The issue of money is made more pressing because activities that currently generate income from the land would have to be significantly reduced if not eliminated altogether.

There is now an opportunity to create in northern New England a wildlands of the sort that Waller proposes. Paper companies own millions of acres of forest in a broad belt that runs along the border with Canada from eastern Maine to the Adirondacks in New York State. While some companies have practiced (more or less) sustainable yield forestry, many more essentially have cut with little thought to the future. These companies now own large tracts of forest that will not be commercially harvestable for decades (though they are anything but wastelands—new growth is abundant and makes much of northern New England good habitat for deer and moose: hence the burgeoning population of moose in southern New England, of which we have already spoken). In the language of business, the land has become a "nonperforming asset." Extractive industries like the wood products industry are notoriously short-sighted and, in a word, impatient. No executive in his or her right mind looks favorably on nonperforming assets. So the orders are going out: sell the land. Since 1995, several million acres of northern forest land have been put up for sale. While there is little chance that this huge expanse would be cleared for human settlement, it is distinctly possible that the forest could become pockmarked with vacation enclaves, eventually inviting more intensive development and complete fragmentation of this extensive forest. Preventing this will require large sums of money, money that the affected states' coffers do not contain. Maine, in which the bulk of the northern forest lies, has a paltry $3 million in reserve for land acquisition, and Vermont and New Hampshire have even less.[11] Only the federal government has the resources to insure that the Northern Forest does not become just another resort and condo opportunity. Indeed, the money is there, in the bank, as it were.

Each year the federal government takes in a little less than a billion dollars from offshore oil and gas royalties. The money is dedicated to the Land and Water Conservation Fund which was established to buy land important for its recreational potential or its ecological importance. Unfortunately, the Republican-dominated Congress has generally refused to appropriate money from the fund, preferring instead to keep as much land as possible in private hands. Taking the public out of the market has meant that the field is largely open to speculators only. Given the time it will take for the forest to become marketable, some private buyers almost certainly will cut what they can and run, rather than wait for the trees to grow. This can only mean resorts and intensified recreational developments. Indeed, in the fall of 1998, Plum Creek Timber Company, a corporation known for rapid timber harvests followed by a strategy of subdivision and resale, bought roughly a million acres of northern forest lands in Maine.

Defensive land purchases are not completely out of the question, even with tight hands on congressional purse strings. In early December 1998, a package of state, federal, and private money was put together by the Conservation Fund to buy 300,000 acres of the northern forest, mostly in New York and Vermont. The $76 million deal will set aside roughly a third of the total land with the remaining 200,000 acres to be sold or leased under strict sustained forestry restrictions. This model, should it become more widely adopted, holds real promise in that it balances the desire for preservation with the needs of local residents, the more general need for forest products, as well as the need for recreational access to our forests. But as welcome as the Conservation Fund deal is, these kinds of purchases will be too few. Without a large federal presence, there simply are not enough dollars around to outbid corporations like Plum Creek.

Money is not the only challenge facing people interested in keeping what remains of our forests out of the hands of subdividers and developers. Environmental groups, most notably the Sierra Club as well as New England-based RESTORE: The North Woods, have launched a campaign to set aside 3.2 million acres of forest in Maine as a "Maine Woods National Park." The goal is to establish a "working ecosystem" instead of a working forest, one that would "protect the self-determining processes of a healthy, diverse ecosystem that generates biological diversity and ecological integrity."[12] By now, it should be clear just how much mischief is masked by this sort of thinking. Never mind that no one knows what a healthy or diverse ecosystem is; never mind that we cannot define ecological integrity: Instead of seeking enlightened restrictions on forestry practices, these groups are pushing to transform the north woods from a "working forest" into a destination for eco-tourism. The goal of this sort of environmentalism is to remove humans as participants in ecological processes and make us spectators. But spectators need wood products too, as well as gasoline, highways and airstrips, motels, and towers to make their cell phones operable, to name only a few of the things that eco-tourism brings to places that boast of being wild. It is far from clear that spectatorship is to be preferred over participation.

The central problem with this way of expressing environmental concern is that it undermines social support for stewardship. It makes suspect all attempts to manage resources, no matter how thoughtful or conscientious the attempt may be. Worse, it obfuscates our intimate dependence on nature. The result is self-deception of the sort that Thom Kyker-Snowman spoke of when he expressed frustration with the well-intentioned people who want all rivers restored to their former free-running condition. Absent hydro-generated electricity, we will be driven even further into dependence on fossil and nuclear fuels. By the same token, opposition to logging our forests, when successful, will only mean shifting logging operations to countries where there are weak, or worse, no environmental regulations. We need, instead, to debate how we use, not whether we should use, natural resources.

Holding up the ideal of ecosystems undisturbed by humans as the model of what environmentalists should seek is likely to drive the wedge between us and the natural world even deeper. It is hard to see how the natural world will benefit.

Given our voracious appetites and growing numbers, each time we restrict the uses to which a particular area may be put, we increase pressures on the rest of the landscape. Some years ago, when the federal government closed the nuclear weapons facility at Hanford, Washington, and for the first time openly confronted the fact that fifty years of weapons production had irretrievably contaminated the area for hundreds if not thousands of years to come, an official resignedly suggested that the area be regarded as a "zone of national sacrifice." Were we to set aside appreciable tracts of land, would we then have to exploit more intensively the nonreserved areas, making them, in effect, zones of national sacrifice? Protection and restoration are crucial to a strategy of caring for the environment, but they cannot be the only ways we display regard for the environment. And given the inevitable trade-offs, it is by no means clear that these expressions of environmentalism, however uplifting they may seem, ought to occupy as much of the stage as they do now.

More than practicalities and trade-offs are involved. In effect, the environmental movement is in danger of being hijacked by people who believe that the only good environment is one rid of human influence and the only true environmentalism is an environmentalism that tries to keep things in their natural state. Elevating "curatorship" over stewardship has the potential of squandering the public's support for environmentalism. Ironically, it also means that both "Wise Users" and "No Users" wind up disparaging efforts to manage natural resources consciously for the common good. More and more of our environment is held hostage precisely in this way.

Back from the Brink

We do not have to let ourselves be captive of either "Wise Use" or "No Use." Reckoning with our many losses, we ought to continue to develop and refine our capacity to reclaim and reinvigorate habitats and, where appropriate, reintroduce locally extinct species of wildlife. But we should be careful not to fool ourselves into thinking that we are in fact putting Humpty Dumpty together again. Whatever things were like before Europeans arrived, it is certain that we will never recapture that condition. We can replicate certain features of that world, but we will never be able to reproduce the biotic circumstances in which nature functioned five hundred years ago. Too many of the variables involved have been irrevocably altered. Some plants and animals are more numerous now than they were on the eve of European settlement, among them white-tailed deer; others are less numerous; and some have vanished altogether. Moreover, the list of new entrants, so-called exotics, grows steadily as commerce and travel link local environments to one another just as surely as our alliances and trading patterns link humans as never before. Even if all new development, logging, fishing, and hunting were now to be stopped, the cumulative impacts of our activities over the past several centuries has significantly changed the mix, the raw ingredients, and thus the course that nature would take left to itself. Even if we had all the pieces, our assembling them would be more akin to the random patterns of a kaleidoscope than anything approaching a replica of the past.

Again, the Quabbin is instructive. Even with the hunt, there are almost certainly more deer on the Quabbin Reservation than there were when the first English settlers arrived in the Swift River Valley. This means that the dynamic interaction between flora and fauna is now different from what it was then. Moreover, the huge body of water we have created there affects everything else—from micro-climates on up. Similarly, it is likely that because of our recently arrived at (and still precarious) solicitude for wolves, there will be more wolves howling in Yellowstone than ever serenaded the indigenous peoples who lived in and moved through the area.* More wolves, a tiny fraction of the former bison herds, thousands of head of livestock around the park,

---

* Daniel Botkin's careful reading of the journals of Lewis and Clark reveals that the expedition encountered wolves very infrequently. Of course it is impossible to know for sure how large the wolf population was two hundred years ago. Nationwide, there were obviously many more wolves and they were found in every state (or what were to become states). But that is not the same as having concentrations of wolves in a few areas. See Botkin, *Our Natural History*.

and millions of tourists each year might all add up to an "intact eco-system" in some peoples' eyes, but it surely bares scant resemblance to what things were like two hundred or more years ago. And even though we will never know what the Quabbin or Yellowstone would look like now had we never set foot in either place, we surely know enough to know that the course of natural history in these places has been funda-mentally altered by our presence.

This is not to disparage all the good work being done to restore habitats. There are many promising efforts underway to boost the recuperative capacities of nature. The *New York Times* science writer William K. Stevens has chronicled the efforts of a group of biologists and environmental activists, marching under the banner of restoration, who are intent on bringing back as much of the original prairie as cur-rent land use can accommodate.[13] Two other science writers, Stephen Budiansky and Gregg Easterbrook, have made even stronger cases for what can be done when the recuperative powers of nature are harnessed to science and careful environmental and resource management tech-nologies.[14] Budiansky and Easterbrook have been denounced in most of the environmental press, a press that seems resolutely committed to juxtaposing man-made travesties to unspoiled natural wonders, as though there can be nothing of worth in between the two.

The idea of restoration is not exactly a new one, though the range and ambition of many recent restorative efforts have given restoration a new cachet. The restoration of wildlife populations has, for example, been a goal of the state and federal agencies setting wildlife policies since the passage of the landmark Lacey Act in 1900. At first, efforts were directed almost entirely at reclaiming habitat and restoring wild-life species of interest to men and women who fish and hunt. Indeed, protecting game species from overexploitation and replenishing the stocks of those species that had been overharvested was one of the main motivations for the environmental policies of Theodore Roosevelt and the early promoters of conservation. Following in this tradition, state and federal wildlife agencies began concerted efforts in the 1960s to restore wild turkeys to much of their former range, and in most of these areas turkey populations are now robust, so robust, in fact, that in some areas people are beginning to complain about nuisance turkeys. Migratory waterfowl numbers similarly had been heading precipitously downward in the middle decades of this century, largely because agricultural practices were destroying the wetlands the birds needed for nesting. But government as well as private efforts have gone far to slowing and in some areas reversing the loss of wetlands, and as a result, the populations of most waterfowl species are at least stable and some have grown. Indeed, the rebound of one bird, the snow goose, has been so marked that it is now threatening to displace many other species of geese and ducks. Efforts to boost the white-tailed deer population have also been embarrassingly successful. Management of habitat and, in some cases, reintroductions have also spurred the spread of beaver and moose over more of their former range. In the East black bear populations are climbing, and in the West there are signs that the grizzly may be staging at least a modest comeback.

Game animals are no longer the only creatures whose numbers and range are deliberately being augmented. Spurred by the Endangered Species Act, nongame species have approached their rightful place in public wildlife policy. The bald eagle and the peregrine falcon have been brought back from the brink, as have a number of other species, plant and animal. Our air and water are cleaner now than they have been in decades. While some major fish stocks are at or near crisis levels, others, like the striped bass, have rebounded vigorously as a result of concerted efforts, both public and private—perhaps, one hopes, an indication that other fish species might also recover if we act appropriately.

Though these and similar efforts are encouraging, no one should imagine that the cumulative effects of the last ten thousand years of human appropriation of nature can be undone. Many life forms have become extinct by virtue of our adaptive success, and it is clear that even with the best of intentions, and an unimaginably concerted effort, many more species will disappear by our hand. These losses will lend urgency to the desire for protection and recovery. Herein lies the danger—if we are too indiscriminate in our efforts, we will surely exhaust ourselves by trying to do more than is humanly possible; if we are too narrowly selective, if we focus our efforts only on a few charismatic species and inspiring habitats, we run the risk of making things worse.

Hubris can be a source of grief, whether it leads us to push creatures away or to bring them back. Right now, public perception clearly favors bringing the wild, in most if not all its forms, back. How long this will be so remains an open question.

The reason for concern is straightforward. Our current embrace of the wild is laying the basis for myriad conflicts between the wild and humans. We have already noted some points of nagging friction: white-tailed deer, of course, but also "nuisance geese" and soon-to-be-nuisance turkeys as well as beaver.* This list now is much longer and is growing steadily. We can add coyotes in New England, a non-native to the region, as well as black bear and moose. Mountain lions in the West are destined to become a problem, especially if other states follow California's lead in banning the hunting or trapping of the big cats. The problem is parallel to the problem of wolf in Yellowstone: we have deliberately or unintentionally been promoting wildlife in prime habitat. But good habitat, as we have seen with wolves, serves as a seed bed for animals who will have to disperse or have their numbers kept in check—or the habitat will cease being good. However balanced and harmonious people imagine wild nature to be, it is neither balanced nor harmonious when city parks, back yards and median strips become the adopted homes of animals sent packing by biological necessity.

Belief in the harmony of nature runs deep and is clearly one of the most powerful metaphors governing our sense of the natural world. As we have seen, critics of the MDC faulted the agency for the "deer problem." Nature would have evened things out and everything would be fine if only we had left things alone. This view of nature makes confronting the reality of our interactions with the wild all the more troublesome precisely because it threatens to cripple policies and practices designed to reduce friction between the wild and humans. We encroach on the habitat of wildlife and they, in turn, encroach on our

---

*Richard Nelson has provided us with a comprehensive analysis of what might be called "the moral career of the white-tailed deer" in *Heart and Blood: Living With Deer in America*. As cities and park authorities grapple with wildlife problems, a new industry is emerging: trappers, sharp-shooters, and specially trained archers whose service is to rid an area of its troublesome animals. For a recent news story about one such operation, see Andrew C. Revkin, "Coming to the Suburbs: A Hit Squad for Deer," *The New York Times*, November 30, 1998, A1, A25.

turf. It would be nice to adopt a live-and-let-live position, but that turns out to be far more easily said than done. Such wishful thinking will only insure the increasing frequency of negative encounters. Over the long run, as these negative encounters multiply, we might well witness an abrupt return to fearing and loathing the wild. Even if we and wildlife are spared that outcome, it is almost certain that the consensus that now exists about the environment will be shattered as communities do battle over how to cope with beaver, geese, coyotes, and all the other critters who are lighting up police switchboards across the country.

We are on a collision course with wildlife. This has been so for roughly the past ten thousand years, the approximate moment when Homo sapiens began to turn from foraging to agriculture. We went from passing among wild creatures to displacing them to make room for our domesticated plants and animals. Even with the best will, it is clear that this collision is unavoidable. The best we are likely to be able to do is to blunt the impact here and there. For those species that can adapt to the environments we create and those that thrive on disturbance, the prospects are not necessarily bad, especially if they can count on our continued solicitude. For others, however, the future does not look bright and I dare say there is little more than hand-wringing that we can realistically offer. Those who think that we are headed for some sort of ecological crash may be right, though I suspect visions of apocalypse are overwrought. If they are right, however, it seems unlikely that we can alter our course quickly and radically enough to avert disaster.

More likely, our choice is not between ecological collapse or learning to live in harmony with nature, but rather between trusting to nature to work things out or accepting our fate as stewards. As stewards, we should endeavor to sustain as richly diverse a biotic neighborhood as we can—for aesthetic, ethical, and scientific reasons. But we also must accept the fact that this entails a heavy dose of management. Trees will be harvested and animal populations will need to be controlled. If we attempt to adopt a "live and let live" stance, attitudes will harden and cleavages will deepen. Just listen to the people who own lakefront property on lakes afflicted with a resident goose population or folks whose subdivision was built near a wetland that is now colonized by beaver. Nature will not resolve such conflicts as these because, as far as nature is concerned, there are no conflicts to resolve. These are our conflicts, and it is up to us, and us alone, resolve them.

We can do this only by being active stewards, not spectators. We have the rich variety of wildlife we currently enjoy precisely because we intervened actively in behalf of wildlife, manipulating their numbers when one species became too numerous, throwing a blanket of protection over species when their numbers got precariously low. For all our mistakes, both of omission and commission, there simply is no other way of proceeding. For better or worse, we are in charge. If we abdicate, we are likely to hand over to the generations to come a much less richly diverse planet than the one we now have. We'd be fools to put the fate of the environment, at this late date, in the lap of nature.

Appendix

Chart 1. Support for Environmental Spending

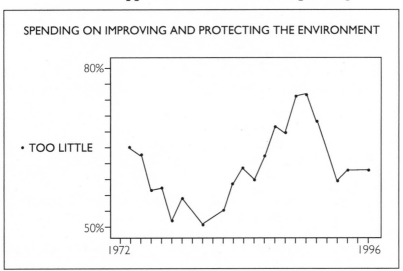

SPENDING ON IMPROVING AND PROTECTING THE ENVIRONMENT

Source: National Opinion Research Center, General Social Survey, 1973–1996

Chart 2. Environmental Protection

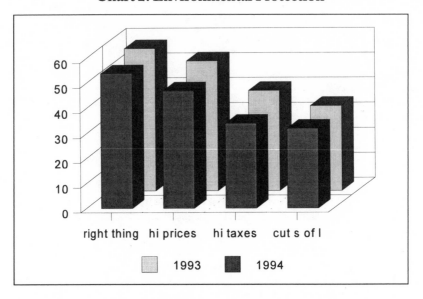

Right thing: "I do what is right for the environment even when it costs more money
      or takes up more time."
Hi prices: "How willing would you be to pay much higher prices in order to protect
      the environment?"
Hi taxes: "And how willing would you be to pay much higher taxes in order to pro-
      tect the environment?"
Cut s of l: "And how willing would you be to accept cuts in your standard of living
      in order to protect the environment?"

Source: National Opinion Research Center, General Social Survey, 1993–1994

Chart 3. Environmental Regulation

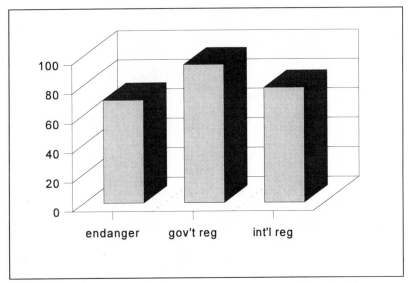

Endanger: "Do you agree or disagree with this statement: Natural environments that support scarce or endangered species should be left alone, no matter how great the economic benefits to your community from developing them commercially might be."

Gov't reg: "On the whole, do you think it should ("agree") or should not be the government's responsibility to impose strict laws to make industry do less damage to the environment?"

Int'l reg: "Do you agree or disagree with this statement: For certain problems, like environmental pollution, international bodies should have the right to enforce solutions."

Source: National Opinion Research Center, General Social Survey, 1996

Chart 4. Man vs. Nature

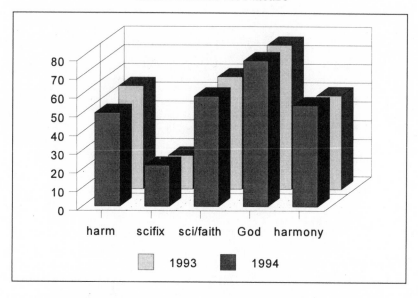

Harm: "Almost everything we do in modern life harms the environment."
Scifix: "Modern science will solve our environmental problems with little change to
       our way of life."
Sci/faith: "We believe too often in science and not enough in feelings or faith."
God: "Human beings should respect nature because it was created by God."
Harmony: "Nature would be at peace and harmony if only humans would leave it
       alone."

Source: National Opinion Research Center, General Social Survey, 1993, 1994

Chart 5. Environmental Attitudes

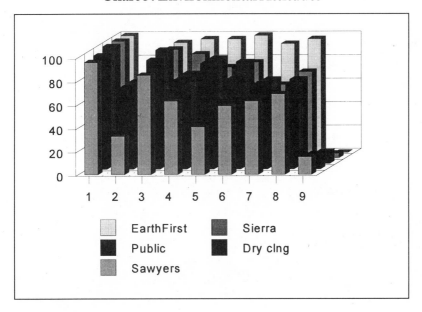

1. We have a moral duty to leave the earth in as good or better shape than we found it.
2. Nature is inherently beautiful. When we see ugliness in the environment, it's caused by humans.
3. Nature may be resilient, but it can only absorb so much damage.
4. Nature has complex interdependencies. Any human meddling will cause a chain reaction with unanticipated effects.
5. Preventing species extinction should be our highest environmental priority. Once an animal or plant species becomes extinct, it is gone forever.
6. All species have a right to evolve without human interference. If extinction is going to happen, it should happen naturally, not through human actions.
7. Humans are ripping up nature, feeling that they can do a better job of managing the earth than the natural system can.
8. Humans should recognize they are part of nature and shouldn't try to control or manipulate it.
9. We shouldn't be too worried about environmental damage. Technology is developing so fast that, in the future, people will be able to repair most of the environmental damage that has been done.

Source: Kempton, et al., *Environmental Values in American Culture*

Notes

1. What's Wild

1. Thomas Conuel, *Quabbin: The Accidental Wilderness*, Amherst: University of Massachusetts Press, revised edition (1990).

2. Terry Tempest Williams, *Refuge: An Unnatural History of Family and Place*, New York: Pantheon Books (1991), 92.

3. Bill McKibben, *The End of Nature*, New York: Anchor Books (1990).

4. Thoreau's views are steadily gaining new and wider attention. Interest will be even further stimulated by the recent publication of some of his heretofore unpublished writing. See Henry David Thoreau, *Faith in a Seed*, Bradley P. Dean, editor, Washington, DC: Island Press (1993).

5. Quoted in Roderick Frazier Nash, *The Rights of Nature: A History of Environmental Ethics*, Madison: University of Wisconsin Press (1989), 37.

6. Michael Pollan, *Second Nature: A Gardner's Education*, New York: Atlantic Monthly Press (1991), 37.

7. Aldo Leopold, *A Sand County Almanac*, New York: Oxford University Press (1949), 7.

8. Quoted in Alston Chase, *Playing God in Yellowstone: The Destruction of America's First National Park*, New York: Harcourt Brace Jovanovich (1987), 26.

4. Sport, Management, or Murder

1. For a lively account of this conflict see Michael Zuckerman, "Pilgrims in the Wilderness: Community, Modernity, and the Maypole at Merry Mount," *The New England Quarterly* 1:2 (June 1977): 255–275.

2. James Tober, *Who Owns the Wildlife? The Political Economy of Conservation in Nineteenth-Century America*, Westport, CT: Greenwood (1981).

3. William Cronon, *Changes in the Land: Indians, Colonists, and the Ecology of New England*, New York: Hill and Wang (1983), 101.

4. First published in Spain in 1942, *Mediations on Hunting* has been translated and reissued many times. The edition from which the quotes below are taken is Howard B. Wescott's translation, published by Charles Scribner's Sons, 1985.

5. This quote is taken from the Modern Library edition of Thoreau's classic, *Walden. Walden & Other Writings of Henry David Thoreau*, Brooks Atkinson, editor, New York: Modern Library (1965), 254.

5. Nature's Rights and Human Responsibility

1. V. C. Wynne-Edwards, "Self-Regulating Systems in Populations of Animals," in *The Subversive Science: Essays Toward an Ecology of Man*, Paul Shepard and Daniel McKinley, editors, Boston: Houghton Mifflin Company (1969), 99–111.

2. "Tragedy of the Commons" originally appeared in *Science* 162 (1968): 1243–48 but has been reprinted in a number of anthologies. Hardin and coeditor John Baden assembled a series of essays elaborating on the original essay in the book *Managing the Commons*, San Francisco: W. H. Freeman (1977).

7. Stewards or Curators? Caring for Nature

1. Wayne F. MacCallum, "Deer are Changing Suburban Attitudes," *Massachusetts Wildlife*, no. 2, 1998, 2–3.

2. Scott Russell Sanders, *The Paradise of Bombs* (Boston: Beacon Press, 1987); Robert Sullivan, *The Meadowlands: Wilderness Adventures at the Edge of a City* (New York: Scribners, 1998).

3. Dire warnings abound. Among the more notable of recent work in this vein, see Niles Eldredge, *Life in the Balance: Humanity and the Biodiversity Crisis* (Princeton, N.J.: Princeton University Press, 1998); Steven Kellert, *The Value of Life: Biological Diversity and Human Society* (Washington, D.C.: Island Press, 1996); and Thomas Berry, *The Dream of the Earth* (San Francisco: Sierra Club Books, 1988). Julian L. Simon's *The Ultimate Resource* (Princeton, N.J.: Princeton University Press, 1981) is a vigorous case against such doom and gloom.

4. Willett Kempton, James S. Boster, and Jennifer A. Hartley, *Environmental Values in American Culture* (Cambridge, MA: Massachusetts Institute of Technology Press, 1995).

5. Paul Schullery, *Searching for Yellowstone: Ecology and Wonder in the Last Wilderness* (Boston: Houghton Mifflin, 1997).

6. Quoted in Jim Robbins, "Yellowstone Reborn," *Audubon*, July-August 1998, 66.

7. See Richard Manning, *Last Stand: Logging, Journalism, and the Case for Humility* (Salt Lake City: Peregrine Smith, 1991); and Nancy Langston, *Forest Dreams, Forest Nightmares: The Paradox of Old Growth in the Inland West* (Seattle: University of Washington Press, 1995).

8. See also Peter Steinhart, The Company of Wolves (New York: Vintage Books, 1996).

9. The symposium was held at the American Museum of Natural History, October 21–23, 1998 and was jointly sponsored by the museum, the Hastings Center, and Defenders of Wildlife.

10. Donald M. Waller, "Getting Back to the Right Nature: A Reply to Cronon's 'The Trouble with Wilderness,'" *The Great New Wilderness Debate*, ed. J. Baird Callicott and Michael P. Nelson (Athens, GA: University of Georgia Press, 1998).

11. *Boston Globe*, July 14, 1998, A1, A11

12. George Wuerthner, "Preserving a Working Ecosystem," *North Woods Vision* 4:1 (April 1996), 6. *North Woods Vision* is the newsletter of RESTORE: The North Woods.

13. William K. Stevens, *Miracle Under the Oaks: The Revival of Nature in America* (New York: Pocket Books, 1995).

14. Stephen Budiansky, *Nature's Keepers: The New Science of Nature Management* (New York: Free Press, 1995); Gregg Easterbrook, *A Moment on Earth* (New York: Viking, 1995).

Sources

I. Published Works Cited

Albanese, Catherine, *Nature Religion in America* (Chicago: University of Chicago Press, 1990).

Benson, Ragnar, *Survival Poaching* (Boulder, CO: Paladin Press, 1980).

Berry, Thomas, *The Dream of the Earth* (San Francisco: Sierra Club Books, 1995).

Botkin, Daniel, *Discordant Harmonies* (New York: Oxford University Press, 1990).

——, Our Natural History: The Lessons of Lewis and Clark (New York: G. P. Putnam's Sons, 1995).

Budiansky, Stephen, *The Covenant of the Wild* (New York: William Morrow, 1992).

——, *Nature's Keepers: The New science of Nature Management* (New York: The Free Press, 1995).

Callicott, J. Baird, ed., *Companion To A Sand County Almanac* (Madison, WI: University of Wisconsin Press, 1987).

Carter, Jimmy, *An Outdoor Journal* (New York: Bantam Books, 1988).

Chase, Alston, *Playing God In Yellowstone: The Destruction of America's First National Park* (New York: Harcourt Brace Jovanovich, 1987).

Conaway, James, "Eastern Wildlife: Bittersweet Success," *National Geographic*, February 1992, 66–89.

Conuel, Thomas, *Quabbin: The Accidental Wilderness*, rev. ed. (Amherst, MA: University of Massachusetts Press, 1990).

Cronon, William, *Changes in the Land: Indians, Colonists, and the Ecology of New England* (New York: Hill and Wang, 1983).

——, "The Trouble with Wilderness; or, Getting Back to the Wrong Nature," *Uncommon Ground: Toward Reinventing Nature*, ed. William Cronon (New York: W. W. Norton, 1995).

Dethier, Vincent G., *The Ecology of a Summer House* (Amherst, MA: University of Massachusetts Press, 1984).

Dickinson, Emily, "What Mystery Pervades a Well," *The Poems of Emily Dickinson*, ed. Thomas H. Johnson, (Cambridge, MA: Harvard University Press, 1955).

Ortega y Gasset, José, *Meditations on Hunting*, Howard B. Wescot, trans. (New York: Charles Scribner's Sons, 1985).

Pollan, Michael, *Second Nature: A Gardener's Education* (New York: Atlantic Monthly Press, 1991).

Regenstein, Lewis G., *Replenish the Earth: A History of Organized Religion's Treatment of Animals & Nature—Including the Bible's Message of Conservation & Kindness to Animals* (New York: Crossroad, 1991).

Russell, Franklin, *The Hunting Animal* (New York: Harper & Row, 1983).

Sanders, Scott Russell, *The Paradise of Bombs* (Boston: Beacon Press, 1987).

Schmitt, Peter J., *Back To Nature: The Arcadian Myth in Urban America* (Baltimore, MD: Johns Hopkins University Press, 1969).

Schullery, Paul, *Searching for Yellowstone: Ecology and Wonder in the Last Wilderness* (Boston: Houghton Miflin, 1997).

Singer, Peter, *Animal Liberation*, 2nd ed. (New York: New York Review of Books, 1990).

Sperling, Susan, *Animal Liberators* (Berkeley, CA: University of California Press, 1988).

Stange, Mary Zeiss, *Woman the Hunter* (Boston: Beacon Press, 1997).

Steinhart, Peter, *The Company of Wolves* (New York: Vintage Books, 1996).

Stevens, William K., *Miracle Under the Oaks: The Revival of Nature in America* (New York: Pocket Books, 1995).

Stone, Christopher D., *Earth and Other Ethics* (New York: Harper & Row, 1987).

Sullivan, Robert, *The Meadowlands: Wilderness Adventures at the Edge of a City* (New York: Scribner, 1998).

Thoreau, Henry David, *Faith In A Seed*, ed. Bradley P. Dean (Washington, DC: Island Press, 1993).

——, *The Maine Woods*, cited in Richard Lebeaux, *Thoreau's Seasons* (Amherst, MA: University of Massachusetts Press, 1984), 180–181.

——, *Walden & Other Writings of Henry David Thoreau*, ed. Brooks Atkinson (New York: Modern Library, 1965).

Tober, James, *Who Owns the Wildlife? The Political Economy of Conservation in Nineteenth-Century America* (Westport, CT: Greenwood, 1981).

U.S Department of Interior, Fish and Wildlife Service and U.S. Department of Commerce, Bureau of the Census, *1980 National Survey of Fishing, Hunting, and Wildlife-Associated Recreation* (Washington, DC: U.S. Government Printing Office, 1982).

Waller, Donald M., "Getting Back to the Right Nature: A Reply to Cronon's 'The Trouble with Wilderness'," *The Great New Wilderness Debate*, ed. J. Baird Callicott and Michael P. Nelson (Athens, GA: University of Georgia Press, 1998).

Williams, Raymond, *Keywords: A Vocabulary of Culture & Society*, rev. ed. (New York: Oxford University Press, 1976).

Williams, Terry Tempest, *Refuge: An Unnatural History of Family and Place* (New York: Pantheon, 1991).

Wynne-Edwards, V. C., "Self-Regulating Systems in Populations of Animals," *The*

Easterbrook, Gregg, *A Moment on Earth* (New York: Viking, 1995).

Ehrenfeld, David, *Beginning Again: People & Nature in the New Millennium* (New York: Oxford University Press, 1993).

Eldredge, Niles, *Life in the Balance: Humanity and the Biodiversity Crisis* (Princeton, N.J.: Princeton University Press, 1998).

Hardin, Garrett, "The Tragedy of the Commons," *Science* 162 (1968): 1243–48.

——, and John Baden, eds., *Managing the Commons* (San Francisco: W. H. Freeman, 1977).

Hillel, Daniel, *Out of the Earth* (New York: Free Press, 1991).

Huxley, Thomas H., *Evolution & Ethics & Other Essays* (New York: Appleton, 1920).

Jackson, Bob, and Bob Norton, "Hunting as a Social Experience," *Deer & Deer Hunting*, November/December 1987, 38–51.

Jasper, James M., and Dorothy Nelkin, *The Animal Rights Crusade: The Growth of a Moral Protest* (New York: Free Press, 1992).

Kellert, Stephen R., *The Value of Life: Biological Diversity and Human Society* (Washington, D.C.: Island Press, 1996).

Kempton, Willett, et al., *Environmental Values in American Culture* (Cambridge, MA: MIT Press, 1995).

Kuhn, Thomas, *The Structure of Scientific Revolutions* (Chicago: University of Chicago Press, 1972).

Langston, Nancy, *Forest Dreams, Forest Nightmares: The Paradox of Old Growth in the Inland West* (Seattle, WA: University of Washington Press, 1995).

Leopold, Aldo, *A Sand County Almanac* (New York: Oxford University Press, 1949).

Lovelock, James, *Gaia: A New Look at Life on Earth* (New York: Oxford University Press, 1979).

McKibben, Bill, *The End of Nature* (New York: Anchor Books, 1990).

McKibben, Hope, *Human and Wild: True Stories of Living Lightly on the Earth* (Boston: Little, Brown and Co., 1995).

McNamee, Thomas, *The Return of the Wolf to Yellowstone* (New York: Henry Holt, 1997).

Manning, Richard, *Last Stand: Logging, Journalism, and the Case for Humility* (Salt Lake City: Peregrine Smith, 1991).

Marks, Stuart A., *Southern Hunting in Black and White: Nature, History, and Ritual in a Carolina Community* (Princeton, NJ: Princeton University Press, 1991).

Marsh, George Perkins, *Man and Nature; Or, Physical Geography as Modified by Human Action* (Cambridge, MA: Harvard University Press, 1965).

Merchant, Carolyn, *Radical Ecology* (New York: Routledge, 1992).

Miller, John M., *Deer Camp: Last Light in the Northeast Kingdom* (Cambridge, MA: MIT Press, 1992).

Mitchell, John, *The Hunt* (New York: Knopf, 1980).

Morris, Desmond, *The Animal Contract* (New York: Warner, 1990).

Nash, Roderick Frazier, *The Rights of Nature: A History of Environmental Ethics* (Madison, WI: University of Wisconsin Press, 1989).

Nelson, Richard, *Heart and Blood: Living With Deer in America* (New York: Knopf, 1997).

Subversive Science: Essays Toward an Ecology of Man, eds. Paul Shepard and Daniel McKinley (Boston: Houghton Mifflin, 1969), 99–111.

Zuckerman, Michael, "Pilgrims in the Wilderness: Community, Modernity, and the Maypole at Merry Mount," *The New England Quarterly* 1:2 (June 1977): 255–275.

II. Unpublished Reports and Proceedings

Carlton, Maggie M., "Literature Review: Water Quality Implications of Converting Forested Watersheds to Principally Herbaceous Cover—Implications to the Quabbin Watershed," submitted to the Metropolitan District Commission, Division of Watershed Management (7 March 1990).

Friends of Quabbin and Metropolitan District Commission, Division of Watershed Management, Quabbin Section, "Quabbin Facts & Figures," n.d.

Metropolitan District Commission, Division of Watershed Management, "1989 Quabbin Forest Regeneration Study," compiled by Thom Kyker-Snowman (7 September 1989).

——, "Deer Browse Impacts on MDC Quabbin Watershed Lands: Answers to Commonly Asked Questions" (25 October 1989).

——, "Draft: Quabbin Reservation White-Tailed Deer Impact Management Plan" (12 July 1990).

——, "MDC Land Management Program: Quabbin Watershed 1995–6 Accomplishments and 1996–7 Program Outline" (11 April 1996).

——, "Quabbin Reservation: White-tailed Deer Impact Management Program, Results of 1997 Program, Recommendations for 1998 Program" (April, 1998).

——, "MDC Land Management Program: Quabbin Watershed 1997–98 Accomplishments and 1998–99 Program Outline" (3 June, 1998).

Natural Resource Management Review Panel, "Report," submitted to the Metropolitan District Commission, Division of Watershed Management (26 May 1989).

O'Connor, Robert, "Recreation and Public Access at Quabbin Reservoir: Conflict in a Commons," Metropolitan District Commission, Division of Watershed Management, n.d.

Public Hearing, Belchertown, MA, 31 July 1990.

Public Hearing, Barre, MA, 7 August 1990.

Public Hearing, Waltham, MA, 8 August 1990.

Public Hearing, Belchertown, MA, 14 May 1991.

Wallace, Floyd, Associates, "Watershed Forest Management Plan: A Technical Review," submitted to the Metropolitan District Commission (April 1989).

III. Personal Interviews (names followed by asterisk indicate the pseudonym used to preserve confidentiality)

Asselin, Ray, 16 July 1991.

Berube, Tom, 9 July 1991.

Boudin, Ron, 23 July 1991.*

Campbell, Elisa, 15 July 1991.
Confessed poacher, 6 August 1991.*
Dodge, Martin, 15 July 1991.*
Edwards, Warren, 12 July 1991.*
Gomes, Peter, 23 July 1991.*
Granby, Bill, 22 July 1991.*
Healy, William, 9 July 1991.
Henderson, Dave, 15 July 1991.*
Hoover, George, 21 August 1991.*
James, Lenore, 13 August 1991.*
James, Steve, 13 August 1991.*
Koski, Charlotte, 19 July 1991.*
Koski, Tom, 19 July 1991.*
Kuznets, Greg, 28 August 1991.*
Kyker-Snowman, Thom, 1 July 1991.
Lewis, Jennifer, 18 July 1991.
Lyons, Paul, 5 August 1991.
McNair, Alexis, 8 July 1991.*
Near, Lorraine, 11 July 1991.*
O'Connor, Robert, 23 August 1991.
Prevost, Laura, 30 July 1991.*
Prevost, Rick, 10 July 1991.*
Read, Clifton, 23 July 1991.
Reading, Dorothy, 3 July 1991.*
Simpson, Liz, 1 July 1991.*
Spencer, Bruce, 9 September 1991.
Thompson, Charlie, 5 September 1991.
Ventura, Jim, 1 July 1991.*
Wagner, Carlton, 21 August 1991.*
Williams, Jack, 3 July 1991.*
Williams, Joan, 3 July 1991.*
Williams, Steve, 19 July 1991.

Index